Bats

Nancy Jennings

BLOOMSBURY WILDLIFE
LONDON · OXFORD · NEW YORK · NEW DELHI · SYDNEY

Dedication

In loving memory of my great friend Kate Barlow (1970–2015), batologist extraordinaire, who used to say 'don't be sad it's over, be glad it happened'.

BLOOMSBURY WILDLIFE
Bloomsbury Publishing Plc
50 Bedford Square, London, WC1B 3DP, UK
29 Earlsfort Terrace, Dublin 2, Ireland

BLOOMSBURY, BLOOMSBURY WILDLIFE and the Diana logo are trademarks of
Bloomsbury Publishing Plc

First published in the United Kingdom, 2018

A catalogue record for this book is available from the British Library

Library of Congress Cataloguing-in-Publication data has been applied for.

ISBN: PB: 978-1-4729-5005-5; ePDF: 978-1-4729-5003-1; ePub: 978-1-4729-5004-8

4 6 8 10 9 7 5

Design by Susan McIntyre
Printed and bound in India by Replika Press Pvt. Ltd.

FSC
www.fsc.org

MIX
Paper from
responsible sources
FSC® C016779

To find out more about our authors and books visit www.bloomsbury.com and sign up for our newsletters

giving
nature
a home

Published under licence from RSPB Sales Limited to raise awareness of the RSPB (charity registration in England
and Wales no 207076 and Scotland no SC037654).

For all licensed products sold by Bloomsbury Publishing Limited, Bloomsbury Publishing Limited will donate a
minimum of 2% from all sales to RSPB Sales Ltd, which gives all its distributable profits through Gift Aid to the
RSPB.

Contents

Meet the Bats

Bats are not the most straightforward animals to get to know. They flit around under cover of darkness, so we rarely see them, and make sounds that are extremely loud but mostly above the range of human hearing, so we cannot hear them. Bats are sometimes viewed, and often portrayed in the media, as weird, creepy, scary or blind, as bloodsucking vampires or fiends, as familiars of wizards or as vital components of witches' magic spells. But take a closer look and you will find that bats are astounding and fascinating creatures.

Bats are incredible, but their abilities mostly remain hidden because they are stealthy and rarely seen. Perhaps that is why a close encounter with a bat is so special, and why I still remember the first time I caught a bat in a net as a research student. I was instantly hooked. The warm, silky, stretchy wings were impressive, but it was the bat's feisty personality that struck me. Looking into the beady eyes of a minuscule bat while holding it gently between your finger and thumb, you might expect it to be terrified, but I received the distinct impression that it was grumpy, outraged and indignant; if it could talk, it would have been ordering me to let it go!

Opposite: The Common Pipistrelle was believed to be the smallest of the 17 bat species that live and breed in the British Isles; however, that position is now contested by the recently discovered Alcathoe Bat.

Left: Greater Horseshoe Bat in typical slow, fluttering flight, using echolocation (see page 16) in an attempt to capture a moth.

Above: Worldwide, the variety of habitats, foods and lifestyles adopted by bats have resulted in impressive diversity. The Spotted Bat is a black and white insect-eater from North America.

Below: Bats can drink while they are flying; this Natterer's Bat is about to take a sip from a pond.

What makes a bat?

The evolution of bats (see page 11) took place over 60–70 million years and was shaped by interactions with other animals and plants. These interactions resulted in the incredible adaptations that bats have today: the ability to fly, to sense their environment, to make the most of the energy they get from their food and to live a nocturnal life.

Alongside their all-important wings, bats have other special adaptations. Their hind feet and legs allow them to hang upside-down when resting; they can take off, land, mate and even give birth from this topsy-turvy perspective. Bat species in cold climates can hibernate to save energy in winter when food is scarce, and all bats can use a great deal of energy to power their flight.

Many bats find their way around by making echolocation calls and listening to the echoes they produce. The complex, high-pitched calls reflect from, or echo off, objects; from these reflections the bats can tell how far away and in what position the objects are, and gauge their texture. They use this special adaptation to detect food in the dark and for orientation. Bats have also evolved ways to locate their young among countless other bats, and to find mates.

Left: Some bats use their sense of smell to find food. Here, a Common Blossom Bat – the smallest species of the Pteropodidae family – is feeding on nectar from a flower in Australia.

Yin and Yang

Biologists today place bats in the order Chiroptera, which is divided into two suborders: Yinpterochiroptera and Yangochiroptera. Yinpterochiroptera comprises the families Pteropodidae (186 species including the flying foxes), Rhinolophidae (horseshoe bats) and five others: Craseonycteridae (one species: the tiny Bumblebee Bat; *Craseonycteris thonglongyai*), Hipposideridae (Old World leaf-nosed bats), Megadermatidae (false vampires), Rhinopomatidae (mouse-tailed bats) and Rhinonycteridae (trident bats). Yangochiroptera comprises all the remaining 14 families of Chiroptera, including the Vespertilionidae (evening bats), the largest family, which has 300 species worldwide and many representatives in Europe.

Above: Davy's Naked-backed Bat of Central America appears naked but there is actually dense fur under the skin (left). The Eastern Red Bat of North America has colourful fur and roosts in trees (right).

Habitats for bats

There are almost 1,400 species of bat in the world, which means that around 25 per cent of all mammal species are bats. In fact, the only group of mammals that is larger is the rodents, which make up about 40 per cent of all mammal species. Of all the bat species that are found worldwide there are 17 that live and breed in the British Isles; all 17 species are included in this book, and we will look at the simple methods that can be used to identify them later (see pages 46–47).

Bats are absent from Antarctica, from very small or isolated islands and from almost all of the Arctic, but they are widespread elsewhere in a diverse array of habitats. Unlike other mammals such as mice, rats and game animals, bats have not been moved around the world much by humans, but their ability to fly has allowed them to colonise many areas, ranging from coastal regions and dense forests to deserts and wetlands. They roost in a variety of places, including caves, cacti, birds' nests and flowers. Some bats even make 'tents' by chewing and bending leaves. Bats make use of human structures for roosting too – mines and tunnels substitute for caves, stone and concrete buildings and bridges provide artificial rock crevices, and wooden buildings have cracks that are used like tree-holes.

Within the habitats they occupy, bats are vital parts of the ecosystem. Bats that eat fruit and nectar perform the crucial jobs of seed dispersal and pollination for the plants that they feed on. Bats that mainly eat insects (such as mosquitoes, midges and moths) have been estimated to take several thousand insects per bat, per night, and therefore have a role in keeping insect numbers down, which is especially helpful when the insects in question are agricultural pests. Without bats and other animals feeding on them, insects would quickly multiply. The minority of bats that don't eat insects, flowers or fruit feed on other animals (invertebrates such as spiders and scorpions, fish, frogs, lizards, birds or small mammals, including other bats); three species of bat feed on the blood of mammals or birds.

Bats worldwide

Bats in the family Pteropodidae, found in the tropics and subtropics of the Old World (Africa, Europe and Asia), mostly eat fruit, especially wild figs, and are also called the fruit bats; the bigger bat species are called flying foxes. You might see these bats in zoos. The largest bat in the world, the Giant Golden-crowned Flying Fox (*Acerodon jubatus*), is a member of the Pteropodidae. This forest-dwelling bat, endemic to the Philippines, is classed as Endangered by the International Union for Conservation of Nature (IUCN), weighs up to 1.2kg (2.6lb; as much as a small rabbit) and has a wingspan of up to 1.7m (5.6ft).

The large flying foxes have excellent vision, but they feed on plant materials (fruit, flowers, nectar and pollen) rather than on fast-moving insects. Very few flying foxes use echolocation, and they do so in a completely different way from other bats, by listening to the echoes of tongue clicks when flying in dark caves. The other, mostly smaller bats have poorer vision and rely heavily on echolocation. The smallest bat, and one of the world's smallest mammals, is the Bumblebee Bat. This species lives in limestone river caves in Thailand and Myanmar, and feeds on tiny insects at night. Adults weigh only 2g (less than 1oz), which is about the same as two small paperclips!

Below: This minute adult Bumblebee Bat was caught in a limestone cave in Thailand. It is the smallest known bat species.

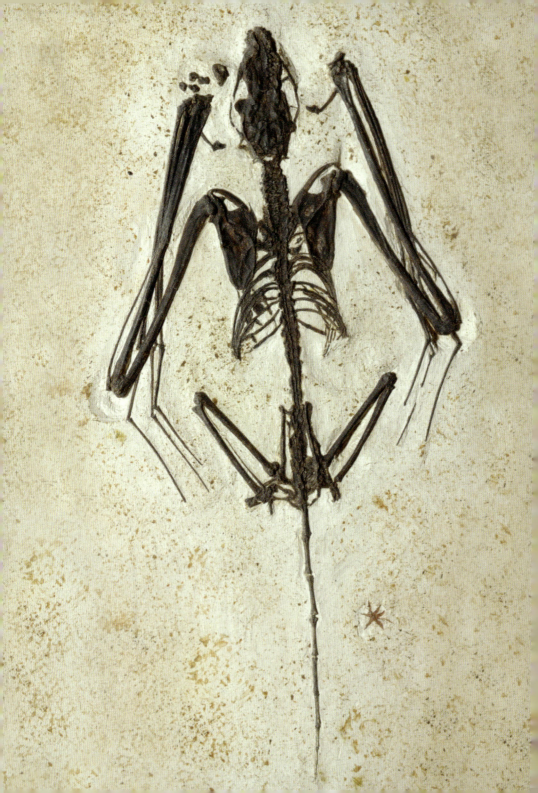

Evolution and Adaptations

Bats, similar to those we know today, have been around for millions of years and scientists believe that they evolved from small insect-eating mammals like shrews. Over time, bats have developed the essential adaptations they need for their lives, which can be grouped into four main areas: flight; echolocation; nocturnality; and energy use, allowing many of them to hibernate and all of them to fly. In this chapter, we'll look at each of these adaptations in turn.

From jumping to flying

Bats evolved 60–70 million years ago, when dinosaurs were coming towards the end of their time (most dinosaurs became extinct around 66 million years ago) and while flowering plants and insects were becoming more common and diverse. It is likely that bats evolved from shrew-like nocturnal insect-eating (insectivorous) mammals that first jumped between the branches of trees, before developing the ability to glide, supported by flaps of skin between their limbs, much like present-day gliding colugos. Bat precursors with some gliding ability would have chased flying insects, so that the next evolutionary step for bats was flapping their wings, resulting, eventually, in powered flight. Over a great many generations the finger and hand bones evolved to support most of the wing, ultimately allowing the highly manoeuvrable flight of bats. The early stages of bat evolution are believed to have taken place in the tropics, where the greatest number of bat species is still found today.

Opposite: This fossil *Icaronycteris index* is one of the earliest known bats; the species has a claw on the index finger that has been lost in modern bats.

Below: The biggest bat in the world, the Giant Golden-crowned Flying Fox, has broad wings suitable for flying in forests, where it finds its food.

Right: The forelimbs of a bird (top), a bat (middle) and a human (bottom). A bird's wing mostly consists of feathers and is not as manoeuvrable as a bat's wing. The bat's finger bones are elongated, but its small thumb is visible just as a claw at the top; the wing connects to the hind foot.

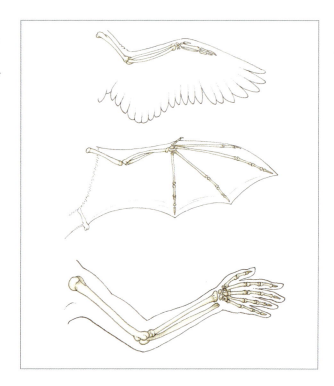

Acrobatic, energetic flight

The wings, evolved from the front legs, are the most obvious features defining bats. Birds fly by flapping their whole front limb (which is equivalent to our arm), but the wing membrane of a bat is supported mainly by the elongated hand and finger bones, so it is really the hand, not the arm, that does the flying. Chiroptera, the scientific name for bats, is derived from the ancient Greek words *cheir* (hand) and *pteron* (wing). A few other mammals (flying squirrels, gliding possums, colugos) are capable of gliding from a high branch to a lower point in the forest by extending special flaps of skin between their front and hind legs, but bats are the only mammals that can genuinely claim powered flight as one of their skills.

A bat's wing is a thin, stretchy double layer of skin, supported by the bones in the limbs; these are mainly the bones we would use in our hands and fingers but also the bones of the fore and hind limbs. In most bats there is

Below: The Sugar Glider is a marsupial gliding possum found in Australia. It cannot fly using its hands like a bat but instead uses the furry flaps of skin between its legs to glide.

also a tail membrane between the hind limb and the tail. The membrane can be used for scooping up insects in flight or for removing them from a leaf, or as a brake. Some bats even flap the membrane up and down to generate lift during take-off. The wings can also be used as insect traps. They have a special cross-sectional shape that allows them to produce lift like the wings of a gliding aircraft. Moving the wings downwards creates the forward thrust needed for powered flight, and the wing shape is changed in the upstroke, so that little drag is created. Bats use their muscles to change the shape of the wings' cross-section while flapping them. It is both the shape and the flapping in combination that generate the upward lift and forward thrust needed to power the bat through the air.

Different species of bat have evolved various flight styles to fit their lifestyles. Some fly slowly and with excellent manoeuvrability so that they can flit around within the crowns of trees, hunting for insects. These bats have relatively big, broad wings and light bodies. Others specialise in hunting the insects found in open spaces or high in the sky; these bats don't need slow flight or manoeuvrability, and therefore their wings tend to be long

Below: This North American bat species, the Spotted Bat, is similar to the long-eared bats found in Europe. In this photo, the tail membrane is visible.

and narrow. Others hunt close to water or at the edges of woodlands. Some have evolved the ability to capture and carry large, heavy insects or fruits, others feed on the wing (while flying), taking smaller, bite-sized insects, and some bats hover when feeding on flowers or gleaning insects from leaves.

The strong flight muscles that power bats' wings have a plentiful blood supply, and bats' hearts are about three times larger than those of small mammals that do not fly. When bats are not airborne their heart rate is like that of other similar-sized mammals, but when they are flying their heart rate increases. A pipistrelle's heart can beat at up to 1,000 times per minute during flight, slowing to 200–450 beats per minute at rest. Hibernating bats can slow their heart rates even more, to around 12 beats per minute.

Flying uses a lot of energy but it is a much faster way for a small animal to move around than crawling or walking, so in terms of the distance that can be covered in a given time flight is energetically good value. Flight also gives bats the ability to search for food over large areas, to catch flying insects, to avoid predators and to access safe, secluded roost sites.

Below: A broad-winged Greater Horseshoe Bat uses its wing tip to scoop up a moth. It will bend its wing to transfer the prey item to its mouth.

Above: Mexican Free-tailed Bats, seen here emerging from a cave roost in Texas, USA, have very long, narrow wings that are suitable for fast flight in open spaces.

Seeing with sound

Bats have a reputation for being blind but they aren't! Echolocating bats use sound as their primary sense to collect information about their environment. Humans and birds mainly use vision; most other mammals, including dogs, have a powerful sense of smell. Not all bats use echolocation; those bats that do produce very loud, high-frequency echolocation calls. The calls move through the air as vibrations, in the form of pressure waves that move air molecules. When the calls reach an object such as a branch, leaf, building or prey, such as insects, they are reflected and some return to the bat as echoes. By listening to the echoes, the bats gain information about their environment.

Echolocation calls are made using the vocal chords in the larynx; they travel out of the bat through its open mouth (in the Vespertilionidae) or its nose (in the Rhinolophidae). Echolocation calls are extremely loud; however, the echoes they create are very faint. To avoid damaging their hearing with the very loud sounds they produce, bats can close their ears by contracting certain muscles in their inner ear many times per second, in time with their calls. Bats often have large or odd-looking ears, mouths and noses – these are adaptations for echolocation and are used to emit, focus and receive sounds. Other animals that use sophisticated echolocation (dolphins, porpoises and toothed whales) produce sonar sounds in their noses and listen to the echoes via a special structure in their lower jaw.

Right: The Harbour Porpoise uses echolocation to find its fish prey in dark, murky water.

How do we describe sounds?

Sounds are made when air (or something else, such as water) is compressed in waves as vibrations caused by a sound source moving through it. The frequency (pitch) is quantified in hertz (Hz), the number of compression cycles per second. High-frequency sounds have more cycles and shorter wavelengths than low-frequency sounds. Bats produce and can hear sounds in the frequency range 8kHz (meaning 8,000 compressions per second) to over 200kHz. The sounds they use are mostly ultrasonic (above the range of human hearing). To hear them, we have to use ultrasound detectors, also known as bat detectors. At the other end of the scale, whales and elephants produce infrasound – sound that is too low-pitched to be heard by us.

The amplitude (relative loudness) of sounds is quantified in decibels or dB. Humans can just about hear sounds of 0dB, and sounds of around 120dB or above are so loud that they are likely to damage human hearing instantly. Bats produce sounds of up to 140dB (measured around 10cm or 4in from the bat), so it is lucky that we can't hear them. If we could, we would need ear protectors when watching flying bats!

Many bat calls are only 2–20 milliseconds (0.002–0.02s) in duration. These very short echolocation calls are produced in series, separated by short gaps. In general, the gap allows time for the bat to detect an echo before producing the next call.

One way to visualise the calls made by bats is with spectrograms, which are plots of frequency against time. From these graphs, the main frequency and the duration of a sound can be read, and loudness is visualised as colour intensity. It is also possible to see how the sound changes over time: does it sweep down in frequency, or remain at a constant frequency?

Almost all the bats found in Europe find their way around by making echolocation calls. These calls are useful for biologists working with bats; they can be recorded and analysed, and are often used to identify the species of bat or to find out more about their habits and behaviours.

Some of the objects that echolocating bats need to avoid bumping into, and some food items (such as fruits and flowers, or animals keeping still to avoid detection) are stationary; others (such as insects in flight) move. Depending on their preferred food, bats may need to be able to detect tiny flies, moths, fruits, flowers, frogs or minute ripples made by fish. They may need to avoid, or land on, branches, leaves, cave walls, buildings, water surfaces or the forest floor. Echolocation is flexible and can meet the challenge of achieving all these tasks. Flying in dense clutter, for instance between the small branches making up the top or crown of a tree, requires different echolocation calls than flying in wide open spaces. Bats

Which came first: flight or echolocation?

Fossils from Germany and the USA suggest that the bats that lived 50 million years ago were not too different from today's bats – they had well-developed wings, ear bones that demonstrate their ability to echolocate and they fed on flying insects. We can make an educated guess about how flight evolved, but scientists have long debated whether flight or echolocation evolved first. Did bats first need to echolocate to be able to fly in darkness without bumping into things? Or did they evolve echolocation once they could fly? Perhaps they did not become completely nocturnal until they had developed sophisticated echolocation.

In 2003, an important piece in the jigsaw puzzle of bat evolution appeared to fall into place. The oldest known bat fossil, estimated to be 52.5 million years old and therefore older than *Icaronycteris index* (see pages 10–11), was discovered in Wyoming, USA. Its scientific name is *Onychonycteris finneyi*. It has wings with claws on all five digits but doesn't have ear bones typical of today's echolocating bats, which suggests that flight came before echolocation. However, when other researchers examined the fossil again, they found structures in the larynx (voice box) that are needed for echolocation, and other early bat fossils do have ears suitable for echolocation. The jury is still out when it comes to deciding which came first, flight or echolocation, but it seems that echolocation has been around for at least 52.5 million years.

Right: The fossilised remains of *Onychonycteris finneyi* – one of the oldest and most complete bat fossils ever found. The detailed anatomy that has been preserved shows how bats evolved. The wing bones are clearly visible and the skull was also preserved (below), however, it was already split from the body when the fossil was discovered.

generally use ultrasound, but within this range they can use very high-frequency sounds with short wavelengths to detect small objects, and less high-frequency sounds with longer wavelengths to detect bigger objects or ones that are further away. The detection of tiny objects requires short wavelengths, as sounds with longer wavelengths may move around the small target without creating an echo.

Echolocation and flight have evolved to function together, and many bats breathe out, making an echolocation call, with every wingbeat. Doing this saves energy and makes both flight and echolocation energetically affordable.

Most bats have reasonable vision, which they can use to find their way around and to locate their prey. Most fruit-eating flying foxes have excellent vision and can see well in bright light as well as in very low light. Frugivorous (fruit-eating) and nectarivorous (nectar-eating) bats also use their sense of smell to find the fruits and flowers that provide their food, and all bats use scent to recognise each other. Some species have special scent glands and tufts of hair that are used to spread their scent. Mothers and babies, as well as adult roost-mates, need to find each other, sometimes in vast colonies consisting of millions of bats.

Above: Many bats use sound waves produced as echolocation calls (yellow) and their echoes (blue) to detect prey, as well as for orientation. The waves are bands of compressed air molecules.

Topsy-turvy life

Birds rest in a position that, from our perspective, is the right way up (with their heads above, or level with, their bodies), but most bats hang upside-down when they are at rest. They temporarily switch to a head-up position to urinate or defecate. Bats can also rest horizontally and many species can take off from the ground. However, their hind limbs are relatively small and rotated, with knees pointed outwards, meaning they cannot support the weight of their upright bodies. So, bats can't perch on a branch in the way that birds can. Hanging from their hind feet allows bats to be ready to fly off quickly, by launching themselves or simply by spreading their wings and dropping. In this way, bats can hang around in places that predators cannot easily reach, like the roofs of caves. Unique adaptations in their feet allow their hind claws to grip even when all the muscles are relaxed; this ensures that bats don't fall off cave walls when they are resting, hibernating or even dead. They don't use energy to hang, so they can truly rest in a position that does not seem restful to us.

Why take the night shift?

For bats, being nocturnal (active at night and at rest during the day) has three significant benefits. First, it is a way to reduce their chances of being eaten, as most birds are diurnal (active during the day and at rest at night) and have evolved to use vision; birds of prey for example, are potential predators of bats. Second, it allows bats to exploit foods that are more easily obtained at night and to avoid competition from insectivorous birds such as Swifts (*Apus apus*) and Swallows (*Hirundo rustica*). Echolocation, unlike vision, doesn't require light; though fewer insects are active at night than during the day, night-flying insects are there for the taking for a flying animal that can detect them. Third, it allows the bats to keep cool, especially in the tropics.

Above: Horseshoe bats rest and hibernate with their wings folded around their bodies, like cloaks.

Hibernation for survival

Bats evolved in the tropics, where food is available year-round and there is no need to hibernate. As they spread into colder areas, bats had to find a way to get through winters when there were very few insects available. Migration was the answer for some, but hibernation was vital for many. During the winter, hibernating bats slow down their heart rate and breathing rate and allow their body temperature to drop in a process called torpor. Hibernation consists of long periods of deep torpor (not sleep). A hibernating bat may take a breath only once every hour, to provide just enough oxygen to the brain and heart to keep it alive. Energy, in the form of fat stored in the body, is conserved in this way so that it lasts until spring. Being small and capable of flight, bats cannot carry much in reserve, and so are living on a knife-edge energy budget.

Below: Barbastelles, seen here hibernating, are rare in the British Isles. However, in other parts of Europe, this species can be found in large groups in caves and bunkers in winter.

During hibernation, bats do arouse (become active) every few days or weeks, to feed, drink, mate, urinate or sometimes just to move around. Arousals cost a lot of energy. Disturbance, if it results in extra arousals, can be very bad for hibernating bats, so they select secluded sites such as mines and caves with a relatively constant, low temperature. Some hibernate in holes in trees or inside cool buildings. Dry places are not suitable for hibernation, as bats would lose too much moisture. If a cave is very damp, bats that hibernate there may be covered with droplets of condensation.

Even bats that don't hibernate sometimes use torpor to save energy. Hibernating species use short periods of torpor on cold days and nights in the spring and autumn to conserve energy when little food is available, or to allow the storage of energy as fat, ready for the winter. The ability to switch between fast, energy-burning flight and slow, energy-conserving torpor is key to being a bat, at least for bats in relatively cool places such as Europe.

Above: Bats prefer damp places for hibernation, so that they don't get dehydrated. Sometimes they become covered with condensation, like this *Myotis* bat.

Bats of the British Isles

Of the almost 1,400 bat species that exist worldwide around 50 species live in Europe, 17 of those live and breed in the British Isles and nine of those are also found in Ireland. All the species in the British Isles feed on insects and some eat other invertebrates, such as spiders, too. British bat species find their way around and hunt mostly by using echolocation, and when the weather is cold, and insects are few and far between, they all hibernate. But British bats are found in different habitats, eat various insects, and differ in size, shape, flight style and diet.

The 17 bat species known to breed regularly in the British Isles belong to two families, the Rhinolophidae (two species of horseshoe bat, which have nose-leaves) and the Vespertilionidae (the remaining 15 species of plain-nosed bats, which have unadorned faces). Globally, these two families have ranges that reach into continental Europe and, in some cases, into North Africa or Asia, extending to China or Japan. So, in the British Isles, most species are on the western edge of their geographic ranges. None of the 17 British species (see page 39) are found in North America, South America, Greenland or Australia. In addition to our resident 17 species, a few others are found very occasionally in isolated locations and are likely to be vagrants.

Opposite: The Noctule is a relatively large bat with strong teeth and sleek, reddish-brown fur.

Left: The Brown Long-eared Bat can be found throughout Europe, including in the British Isles.

Lucky horseshoes?

The Greater Horseshoe Bat and the Lesser Horseshoe Bat are pretty similar except in size. Their horseshoe-shaped nose-leaves (fleshy structures on their nose) make them easy to recognise if you see them close up. Both species tend to hang freely when roosting and wrap their wings around themselves like cloaks. In this position, with wings folded, the Greater Horseshoe Bat is about the size of a pear and the Lesser Horseshoe Bat is about the size of a plum. Both have fluffy brownish to yellowish fur that is paler on the tummy than on the back, and both forage close to their roosts, mostly travelling up to 2.5km (1.6 miles) to feed, very rarely 10–20km (6–12 miles).

The Greater Horseshoe Bat suffered a dramatic decline in the British Isles (by about 90 per cent over approximately 100 years), but numbers have been increasing since the 1990s and there are now estimated to be more than 9,300 individuals. Though it is still a widespread species, occurring in Europe, North Africa and south-west Asia, it has become extinct in the Netherlands, most of Germany and other areas. In the British Isles, Greater Horseshoe Bats hibernate in caves and feed mostly on moths, dung beetles

Below: Around the size of a pear, the Greater Horseshoe Bat tends to hang freely from the roof of a cave, with its wings wrapped around its body (left). In flight, it is slow and fluttering (right).

and cockchafers found over cattle pasture. They are large, robust bats with big teeth. They hunt by flying close to the ground over grassland, and also by perch-hunting: making short flights from a resting place to capture big prey items, which they take back to the perch to consume.

Lesser Horseshoe Bats are much smaller and more delicate, and feed mostly in woodland on mosquitoes, crane flies, gnats, blackflies and midges, all caught and eaten on the wing, sometimes in or near dense foliage. These bats prefer to avoid flying in open spaces. Instead, they follow treelines, hedges and forest edges. Their numbers are increasing in the British Isles.

The wings of both species are short, broad and rounded, allowing very slow, fluttering flight. Both horseshoe bats use echolocation calls that are distinct from those of all the other species in the British Isles: the frequency is very high (around 83kHz for the Greater Horseshoe Bat and typically 112kHz for the Lesser Horseshoe Bat) and remains constant during most of each call, whereas calls of other bats sweep down through a range of frequencies. The calls are also much longer than those of other bats in the British Isles, and the gaps between the calls are much shorter. These special calls are focused by the nose-leaf and have evolved for detecting moths. They are above the range of hearing of most moths.

Below: The Lesser Horseshoe Bat flies slowly and feeds on small flies (left). When roosting, it wraps its wings tightly around its body and is about the size of a plum (right).

The mouse-eared bats

Six similar species of mouse-eared or *Myotis* bat are found in the British Isles; all weigh up to about 10g (0.4oz). Daubenton's Bats are perhaps the most well known and the easiest to identify when they are foraging in their favourite place, just above the water surface of rivers and lakes.

Go to a smooth expanse of water on a summer's evening and you may see Daubenton's Bats zooming around in circles and ovals about 10cm (4in) above the water – often touching the surface when they catch an insect. Other bats come to the water to drink but don't feed in this distinctive way. Many insects live in the water as larvae when they are immature and cannot yet fly. It is when they reach adulthood and emerge that these aquatic insects – midges, mosquitoes, caddisflies, mayflies, lacewings and others – become the main prey of Daubenton's Bats. Compared with other *Myotis* bats, Daubenton's Bats have small ears and large, hairy feet; they catch insects by dragging their feet through the water as they fly over. They roost in tree-holes or crevices in bridges, in nursery colonies of up to a few hundred females. Males roost in smaller groups. In common with other species, Daubenton's Bats often follow fixed flight paths between roosts and feeding areas and prefer to fly along linear features such as hedges, ditches, forest edges, roads or trails.

Below: A Daubenton's Bat capturing insect prey at the water surface by using its large hind feet to trawl through the water.

Brandt's Bats and Whiskered Bats can be almost impossible to tell apart and were considered to be the same species until 1970. Brandt's Bats inhabit damp forests, wooded river valleys and swampy areas, and hunt along tree-lines and hedges. They feed on moths, spiders, earwigs and flies (including crane flies and midges). The non-flying prey are gleaned from the ground or vegetation. Brandt's Bats are agile flyers and use undulating flight when in the open; they also hunt over water but usually higher above the surface than Daubenton's Bats. Whiskered Bats are small and very lively and have frizzy brown fur on their backs. They hunt in open habitats, villages, orchards, gardens, forests and along streams. Their flight is highly manoeuvrable, so they can even enter the crowns of trees. They eat insects such as crane flies, midges, mosquitoes, moths and lacewings, caught mainly on the wing.

Above: Brandt's Bats are agile in flight and can glean insects from surfaces as well as taking them in flight.

Below: Whiskered Bats can fly in open spaces, but their manoeuvrable flight also allows them to fly in dense vegetation.

Above: The recently discovered Alcathoe Bat, a small *Myotis*, is the smallest of the 17 bat species that live and breed in the British Isles.

The Alcathoe Bat, first documented in Greece and Hungary in 2001, was added to the list of bats living and breeding in the British Isles in 2010. It is very closely related to Brandt's Bats and Whiskered Bats. The Alcathoe Bat is the smallest European *Myotis* species and it lives in dense, often wet forests with very large, old trees of many species, where it roosts in tree cavities or behind bark, usually in oak trees. It feeds on the wing, and eats small moths and mosquitoes.

Natterer's Bats live in tree holes, bat boxes and barns; they have long ears and eat a lot of creatures that don't fly or don't fly at night, such as diurnal flies, spiders, woodlice, harvestmen, crane flies and caterpillars. Natterer's Bats have bare, pink faces. They feed in open park-like areas and all types of forest, and often scoop prey from the ground or from vegetation (a method called gleaning). To assist in gleaning prey, Natterer's Bats have a distinctive double row of curved bristles on the edge of their tail membrane. Their flight is very manoeuvrable and slow, allowing them to hover and fly very close to vegetation, as required for gleaning. Natterer's Bats may also feed by landing on the ground and pursuing prey there.

Bechstein's Bats are rare and declining forest dwellers, who prefer old oak and beech forests, where they glean moths, flies, spiders, beetles, lacewings, earwigs, caterpillars and bugs from tree canopies or just above the ground. They don't go far from their roosts and can hover

Left: Natterer's Bats have a distinctive bare, pink face, as well as pink limbs that earned them the nickname 'red-armed bats'.

and fly with great manoeuvrability. Bechstein's Bats have very long ears – of those found in the British Isles, only the long-eared bats have longer ears – and if they are folded forwards, more than half the ear sticks out beyond the end of the muzzle. Bechstein's Bat detects some of its prey by listening for sounds the insects and other invertebrates make as they scuttle around. Though the species is rare today, fossilised remains of Bechstein's Bat are the most common bat fossils at some Neolithic sites in England. It is likely that this species declined when humans cleared the forests for agriculture thousands of years ago.

Below: Bechstein's Bats fly slowly and prefer to feed in forests. This individual's long ears are visible.

Bigger British bats

Noctules, Leisler's Bats and Serotines are comparable in size (relatively big) and sound similar on a bat detector. However, they vary in their flight style, live in different habitats and feed on dissimilar groups of insects.

Noctules have sleek, reddish fur, broad, rounded ears and long, narrow wings that allow them to fly very fast, at over 50km/h (31mph) – most smaller bats fly at about 20–30km/h (12–19mph). The Noctule is the biggest and heaviest species found in the British Isles, but it still weighs a lot less than a House Mouse (*Mus musculus*), which tips the scales at about 40g (1.4oz). With their long, narrow wings, Noctules fly high in the sky, often at 10–50m (32–164ft) above the ground, sometimes to 100m (320ft) or more, making rapid dives to catch insects (flies, bugs, caddisflies, beetles and moths) and not normally flying close to obstacles. They live in or near lowland woodland, preferring to roost in tree-holes (especially old woodpecker holes) in trees on the edges of woods. From early August, males spend time singing from breeding holes in trees to attract females. Noctules in the British Isles are not known to migrate, but those in other parts of Europe travel 1,000km (621 miles) or more to the south-west for the winter.

Below: Also known as the Hairy-armed Bat, the Leisler's Bat is a forest species that often roosts in tree-holes.

Above: Serotines have broad wings and rounded tail membranes, and are able to fly slowly.

Leisler's Bats are similar to Noctules but are browner, smaller and lighter. They eat moths, flies (including a lot of yellow dung flies) and aquatic insects such as caddisflies, and are often found in woodland or over pastures and lakes. Summer roosts are usually in natural tree-holes, such as those caused by lightning or rot, but are also found in buildings. Nursery roosts normally contain 20–50 females, but there is one on record in Ireland with up to 1,000 Leisler's Bats! These bats, like the Noctules, have long, narrow wings and fly fast and high. Unusually, their underfur extends onto the wing and along the forearm. Males attract females with a special display of flying and calling (a songflight) or by singing from roosts.

Serotines have much broader wings than Noctules and Leisler's Bats, so they fly more slowly and are more manoeuvrable in flight; they even occasionally pick up prey from the ground. They hunt in many different habitats, from farmland and villages to forest edges and rivers. Serotines eat mainly beetles and cockchafers. In summer, nursery roosts are mostly in buildings and consist of 10–60 females.

The pipistrelles

If you see a small bat flitting around near a tree or building, it is pretty likely to be a pipistrelle. Until the early 1990s only one species of pipistrelle was known in the British Isles, and it was our most common bat species by far. That species has since been found to consist of two very similar species of bat (referred to as cryptic species), the Common Pipistrelle and the Soprano Pipistrelle – which differ in the frequency of their echolocation calls. A third species, Nathusius' Pipistrelle, has also been found to breed here, though it is much less common. Male Nathusius' Pipistrelles attract females by singing from breeding roosts in autumn; males of the other two species perform songflights.

Common Pipistrelles can be found from southern Finland to north-west Africa and are as much at home in cities and villages as in woodland and around water. Their diet consists mainly of flies, caught on the wing while flying around the edges of trees and near street lamps. Summer nursery colonies consisting of 50–100 female Common Pipistrelles form from May, usually in buildings, and move regularly, sometimes as often as every few days. Their position as the smallest bat species in the British Isles is contested by the recently discovered Alcathoe Bat.

Soprano Pipistrelles are more likely to be found around water than Common Pipistrelles. They particularly like lakes and ponds, but they also feed on flies, mayflies and lacewings close to vegetation along rivers. Aquatic insects dominate their diet. Soprano Pipistrelles tend to form large colonies in buildings in summer, consisting of up to 800+ females. Males, like the males of many other bat species, roost alone or in small groups, and are very hard to find in summer.

Nathusius' Pipistrelles in continental Europe migrate in a south-westerly direction to avoid the very cold winters, moving towards more

Below: Common Pipistrelles roost in buildings, tree-holes and crevices. They catch small insects that they eat on the wing. Each individual can consume up to 3,000 insects in one night!

maritime climates. These tiny bats really are long-distance travellers; the maximum recorded distance an individual has travelled is almost 2,000km (1,243 miles)! During migration, they travel about 30–50km (19–31 miles) per night. A male Nathusius' Pipistrelle ringed at Blagdon Lake, near Bristol, UK, was discovered 596km (370 miles) away in the Netherlands in 2013, providing the first firm evidence that this species can cross the North Sea. More recently, Nathusius' Pipistrelles ringed in Latvia and Lithuania – countries that are both more than 1,400km (869 miles) from the UK – have been caught in the UK. Nathusius' Pipistrelles are reddish brown in colour and use trees as hibernation sites. They eat a lot of aquatic insects (mostly flies, such as midges and mosquitoes, and caddisflies, aphids and alderflies), and like to forage in forests, along rivers, and over floodplains and reed beds.

Above: Soprano Pipistrelles are similar to Common Pipistrelles, but their echolocation calls are at a higher frequency and their faces are pinkish brown.

Left: The Nathusius' Pipistrelle is a long-distance migratory species. The slightly longer fur on its back gives it a shaggy appearance.

The moth-eaters

Above: Barbastelles are true moth specialists, with distinctive broad ears that meet across their forehead.

Below: The broad wings and large tail membrane of the Brown Long-eared Bat enable slow, manoeuvrable flight that often includes steep dives and short glides.

Most insectivorous bats eat moths when they can, but the Barbastelle and the long-eared bats are moth specialists. The rare and declining Barbastelle is distinctive; its big ears have folds on the outside edges, face forward and meet across its forehead. It has long, silky, blackish fur with whitish tips, giving it a frosted appearance. It forages in old forests, gardens and along hedgerows – its agile flight allowing it close access to vegetation – taking almost entirely moths. In summer, Barbastelles often roost under bark, in tree crevices and in bat boxes; in winter they hibernate in deep holes in trees, or in caves, mines, bunkers and ruins.

The Brown Long-eared Bat and the rarer Grey Long-eared Bat are tricky to tell apart even when seen close up. Both have very long ears and are brownish, though the Grey Long-eared Bat has a dark grey muzzle and greyish-brown fur, while the Brown Long-eared Bat has a light brown face, brown fur, and larger thumbs, claws and feet. The slow, fluttering flight of the Brown Long-eared Bat

makes it particularly vulnerable to road traffic accidents; it is the species most often killed on the roads in the British Isles. It is found in orchards, parks and gardens, and feeds in flight and by gleaning prey from vegetation, which involves hovering flight. Both long-eared bat species locate their prey not only by using echolocation but also by listening with their big ears and by watching for movement. They carry large prey items to special perches to eat them; moth wings and other insect remains can be found below them. Brown Long-eared Bats mostly eat moths, as well as some flies, grasshoppers and many non-flying invertebrates such as spiders, earwigs and caterpillars. In the British Isles, the Grey Long-eared Bat is found only in the south. It frequents villages and other areas with a diversity of habitats, and feeds more on flying insects – 70–100 per cent of its diet consists of moths. However, like the Brown Long-eared Bat, it can glean prey items from leaves.

Above: Meadows and woodland edges are good places to look for foraging bats. This Brown Long-eared Bat is using its ears to listen for sounds made by insects or spiders in this meadow's long grass.

Bats resident and breeding in the British Isles

Common name	Scientific name	Weight	Range, notes and IUCN Red List category
Lesser Horseshoe Bat	*Rhinolophus hipposideros*	4–7g (0.1–0.2oz)	South-west England, Wales, west Ireland; Least Concern
Greater Horseshoe Bat	*Rhinolophus ferrumequinum*	18–24g (0.6–0.8oz)	South-west England, south Wales, Channel Islands; Least Concern
Daubenton's Bat	*Myotis daubentonii*	6–10g (0.2–0.4oz)	Most of British Isles including Scotland and Ireland; Least Concern (generally increasing in Europe since 1950)
Brandt's Bat	*Myotis brandtii*	5–7g (0.2oz)	Separated in 1970, found in England and Wales; Whiskered Bat also in south Scotland, Ireland and on the Isle of Man; both Least Concern
Whiskered Bat	*Myotis mystacinus*	4–7g (0.1–0.2oz)	
Alcathoe Bat	*Myotis alcathoe*	4–6g (0.1–0.2oz)	Separated from the Whiskered Bat in 2001, confirmed as resident in the British Isles in 2010, present on Channel Islands; Data Deficient
Natterer's Bat	*Myotis nattereri*	7–10g (0.2–0.4oz)	Most of British Isles including Scotland, Ireland and Channel Islands; Least Concern
Bechstein's Bat	*Myotis bechsteinii*	7–10g (0.2–0.4oz)	South England and south-east Wales; Near Threatened (rare and declining)
Noctule	*Nyctalus noctula*	21–30g (0.7–1.0oz)	England, Wales and south Scotland; Least Concern
Leisler's Bat	*Nyctalus leisleri*	13–18g (0.5–0.6oz)	Ireland, England, east Wales, south Scotland, Isle of Man; Least Concern
Serotine	*Eptesicus serotinus*	18–25g (0.6–0.9oz)	Central and south England, north Wales, Channel Islands; Least Concern
Common Pipistrelle	*Pipistrellus pipistrellus*	3–7g (0.1–0.2oz)	Most of British Isles, including Channel Islands; Soprano Pipistrelle not found in north Scotland and north-west Ireland; separated in 1990s; both Least Concern
Soprano Pipistrelle	*Pipistrellus pygmaeus*	4–7g (0.1–0.2oz)	
Nathusius' Pipistrelle	*Pipistrellus nathusii*	6–10g (0.2–0.4oz)	Confirmed as resident in 1997; migratory; England, Wales, north Ireland, east Ireland, south Scotland, Channel Islands; Least Concern
Barbastelle	*Barbastella barbastellus*	7–10g (0.2–0.4oz)	South England and Wales; Near Threatened (rare and declining)
Brown Long-eared Bat	*Plecotus auritus*	6–9g (0.2–0.3oz)	Most of British Isles, including Ireland and Channel Islands; Least Concern
Grey Long-eared Bat	*Plecotus austriacus*	6–10g (0.2–0.4oz)	South England, south Wales and Channel Islands; separated from Brown Long-eared Bat in 1960; Least Concern

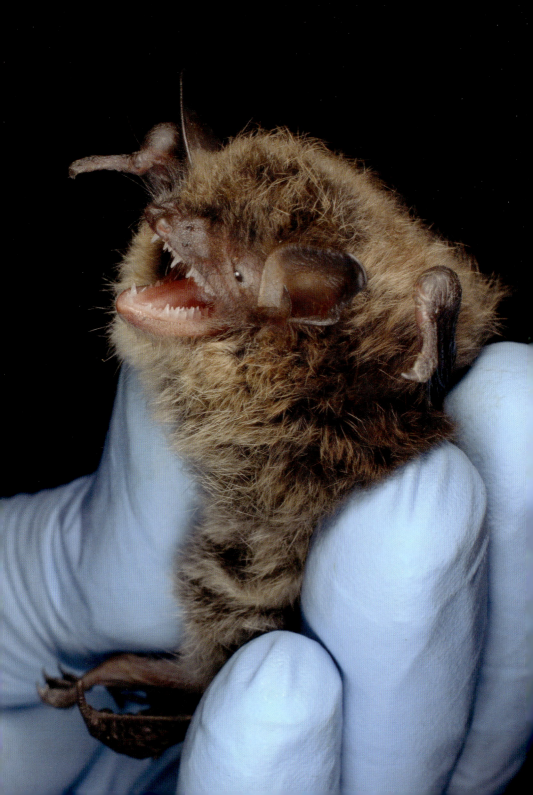

Identifying Bats

Bats are hard to see in the darkness and you can't hear their ultrasonic echolocation calls, so they can be very challenging to identify in flight. Using bat detectors to make their echolocation calls audible can help, and if you get a good view of a flying bat, the wing and tail shape, as well as the bat's flight style, may give you some clues. A bat in the hand is more straightforward to identify, but you may need expert knowledge, an identification guide and some equipment.

All the bats in the British Isles are small and brownish, and prefer to show themselves only at dusk or in darkness, when binoculars are not much use. Birds tend to be easier to see, as they may perch on treetops and fences, sing or come to bird tables. Partly because of their stealthy behaviour, new and cryptic species of bat are still being discovered. Bats are secretive and tricky, but whether you have a bat in the hand, see a bat flying, find a roost or see evidence of a bat, you can at least have a go at identifying it. It is worth bearing in mind your location and the geographical ranges of the species: for example, if you are in Scotland or Ireland, you have fewer species to consider than if you are in south-west England where all 17 species are present (see page 39).

Opposite: A Brandt's Bat caught by the Wiltshire Bat Group during an autumn swarming survey near Box, Wiltshire, UK.

Left: Daubenton's Bats have smaller ears than their close relatives the other *Myotis* bats.

Tiny but feisty

Right: A Common Pipistrelle caught during a survey by the Wiltshire Bat Group.

Below: Noctules are much bigger and redder than pipistrelles. When handled, they are more likely to remain calm, whereas little pipistrelles tend to wriggle.

Seeing a live bat close up for the first time is a revelation to most people. You might be lucky enough to meet a bat rehabilitator who cares for injured bats, and get a chance to see a bat in the hand. If you do, it is likely to be a pipistrelle, as the two most common species in the British Isles are the Common Pipistrelle and the Soprano Pipistrelle. Both species are endearing but also tend to be feisty (when viewed in the hand, they will wriggle and bite their carers' gloves), with what can only be described as forceful personalities. Pipistrelles have tiny, beady black eyes (smaller than pin-heads) that are bright and lively, very sharp teeth (like little white needle points) and amazingly soft fur. When its wings are closed, a pipistrelle fits comfortably into a matchbox. When they're spread out, however, their wings are much bigger than their body and are made of soft, silky-smooth and incredibly stretchy, warm skin. Other species in the British Isles tend to be more laid-back but are not much bigger than the pipistrelles. The sleek Noctule is the biggest species, but with wings folded even it still fits easily into the palm of an adult's hand. Noctules weigh 21–30g (0.7–1.0oz), which is about as much as two mice, or five pipistrelles.

A bat in the hand

Handling live bats is not recommended and is illegal in the UK unless you are a trained licence-holder or rehabilitator. This is because handling may cause distress to the bat and may expose the handler to a type of rabies, though the risk of contracting rabies from a bat in the British Isles is low. However, if you find a live bat that is grounded or injured, you may want to try to help it, and handling for rehabilitation and release is permitted. The best thing to do is to move it into a bag or box using a cloth or while wearing gloves. If you have nothing else to hand, take your sock off and get the bat into it, then hold the top of the sock gently closed. The bat will probably hang quietly upside down (or right way up for a bat), so that its teeth are well away from your hand. When possible, transfer the bat and the sock to a small box and provide a tiny container, such as a milk bottle cap, with water. If you are in the UK, call the Bat Conservation Trust's National Bat Helpline (see page 125) and you will be put in touch with a volunteer bat rehabilitator.

If you find a dead bat in the UK, it is illegal to keep it in your possession but you are allowed to take a good look. Handle the bat with gloves if possible, and report it to the Bat Conservation Trust via the National Bat Helpline. To identify it with certainty as a given species, you would

Left: A rescued bat in a cardboard box is taken to North Devon Bat Care, a rescue centre in the UK, for rehabilitation.

Right: A digital caliper is used to measure the forearm length of a bat. Ideally, gloves should be worn when handling bats.

Opposite: These biologists are examining a recently discovered bat in the Caucasus, Georgia.

need an identification key and an accurate set of scales (though dead bats may have dried out and weigh less than live ones). You would also need a ruler or caliper to measure various parts of the body, and a magnifying glass or hand lens to examine things like teeth, veins in the wing and the penis if the bat is male.

Even without any of this special equipment, you will be able to narrow the species down a little (see pages 46–47).

Sexing a bat is easy: male bats have large, very obvious genitals. On females, the nipples are tucked away under the wings, but if the female was feeding young when she died you might see them if you gently part the fur. Young bats can be especially confusing to identify; the measurements given overleaf are for adults, so if you find a bat that doesn't seem to fit any species, it might be a baby.

Below: On male bats, the penis is usually quite obvious, as in this male Geoffroy's Bat (left). Females, like this Greater Horseshoe Bat, may have visible nipples (right).

IDENTIFYING BATS

Identifying the bats of the British Isles

Start by looking at the face: does the bat have a plain nose or a horseshoe-shaped nose-leaf? If it has a nose-leaf, it will be easy to use its weight or size to identify it as a Greater Horseshoe Bat (over 15g or 0.5oz) or a Lesser Horseshoe Bat (under 10g or 0.4oz).

If it has a plain nose, it could be any of the 15 Vespertilionid species, but some are pretty easy to identify.

The complex horseshoe-shaped nose-leaf is a distinguishing feature of Greater and Lesser Horseshoe Bats.

The nose of a Vespertilionid species (for example a pipistrelle) is much plainer.

If the ears are joined over the forehead and longer than 2cm (0.8in), it is a long-eared bat. The ears may be curled back and tucked away. The Brown and Grey Long-eared Bats are tricky species to separate, but the overall colour gives a clue. If the ears are joined, but shorter than 2cm (0.8in), and the bat has blackish fur, it is a Barbastelle.

The ears of a long-eared bat are nearly as long as its body, and have folds and ridges along the outer edges.

When at rest, long-eared bats tuck their ears back so only the pointed tragus is visible.

Barbastelles have broad ears that meet over the forehead.

If the ears are separated over the forehead, it gets a bit trickier, but if the bat weighs more than about 13g (0.5oz) it is likely to be one of three species: a Noctule (over 20g or 0.7oz, sleek reddish fur, narrow wings), a Leisler's Bat (13–18g or 0.5–0.6oz, sleek brown fur, narrow wings) or a Serotine (18–25g or 0.6–0.9oz, fluffy brown fur, broad wings).

Noctules can often be identified by their sleek, reddish-brown fur.

Leisler's Bats have longer fur around the shoulders and upper back than on the rest of their bodies.

Serotines have a distinctly black face and ears.

In smaller bats, look for the tragus, which is a fleshy structure sticking out inside the ear; if it is short, blunt and stubby, the bat is a pipistrelle, if it is long, pointed and narrow, it is a *Myotis* bat.

The tragus of a Common Pipistrelle is short, blunt and stubby.

The tragus of a Daubenton's Bat (or other *Myotis* bat) is long, pointed and narrow.

The three little pipistrelles are hard to separate, though the Nathusius' Pipistrelle's fur is more rough-looking because the individual hairs are lighter at the tip. To separate the Common and Soprano Pipistrelles you really need a live, flying bat and a bat detector, as their echolocation calls differ (and this is the easiest way to tell them apart). Common Pipistrelles tend to have blacker faces, but it is hard to see the difference even if you have one of each species in your hand.

Nathusius' Pipistrelles have reddish-brown fur; the ears and face are usually dark.

The Common Pipistrelle is medium to dark brown with a blackish face.

Soprano Pipistrelles, with pinker faces, can be separated from Common Pipistrelles by their echolocation calls.

We are left with the six *Myotis* species. If the bat has long ears (over 2cm or 0.8in, sticking out beyond the muzzle if folded forward) it is a Bechstein's Bat. If it has big hairy feet, it is a Daubenton's Bat. If it has a row of bristles on the edge of its tail membrane, it is a Natterer's Bat. If it is one of the remaining three species, you will need a magnifying glass and a detailed key. You could just put it down as a Brandt's/Whiskered/Alcathoe Bat.

Bechstein's Bats have relatively long ears

The large hairy feet of Daubenton's Bats can be used to distinguish them from other *Myotis* species.

The stiff bristles along the trailing edge of the tail membrane are a distinguishing feature of Natterer's Bats.

Bats in flight

Each species of bat has a particular flight style, so you can at least make an educated guess about its identity if you get a reasonable view of a flying bat. If you are lucky, you might be able to see the shape of the wings, ears or tail membrane.

If you see a small bat flying fast and in a chaotic way, around the edges of things like trees, buildings or near a street lamp, it is very likely to be a pipistrelle – they are common and widespread. A small bat that is flying around inside the crowns of trees or shrubs might be a Whiskered Bat, one of the other small *Myotis* species (a Brandt's Bat or Alcathoe Bat), or a Lesser Horseshoe Bat. Bechstein's Bats fly slowly, close to vegetation. A bat making circles or ovals a few centimetres over the surface of a smooth body of water such as a pond, lake or river, and touching the surface every few minutes, is likely to be a Daubenton's Bat. Small bats flying slowly and sometimes hovering could also be long-eared bats; you might see the ears pointing forwards. A large bat flying with slow, fluttering flight and broad wings, close to the ground or to vegetation, could be a rare Greater Horseshoe Bat. A broad-winged Serotine is likely to fly slowly, too, but slightly further away from obstacles. But if you see a large narrow-winged bat flying very high in the sky, in a straight line, looking a bit like a Swift or a Swallow, it could very well be a Noctule, or perhaps a Leisler's Bat.

These identification methods are by no means certain and, unfortunately, you can never check them, though you may get a clearer picture if you have a bat detector. If bats are out in the early evening, as they sometimes are after a rainy night when they are particularly hungry, you could try taking photos. It is difficult, but you may be able to zoom in and compare the shape of the wing and tail membrane to a photo or a flight silhouette (see opposite). Don't disturb bats by using flash photography, or by taking photographs in or near a roost.

Above: The shape of the wings and tail membrane of flying bats can sometimes be used to aid identification. Here are the flight silhouettes, approximately to scale, of: (a) the broad-winged Greater Horseshoe Bat (the Lesser Horseshoe Bat is similar but smaller); (b) a *Myotis* bat, in this case Bechstein's Bat with relatively long ears; (c) the Noctule, with long, narrow wings and a wedge-shaped tail membrane; (d) the Common Pipistrelle (the Soprano Pipistrelle and Nathusius' Pipistrelle are similar); (e) the Serotine, with broad wings and a rounded tail membrane; and (f) a long-eared bat (Brown and Grey Long-eared Bats are very similar).

Bats in a roost

If, at dawn or dusk, you see a stream of bats flying into or out of a building, tree-hole or bat box, you have discovered a bat roost. Don't try to look inside, as this will disturb the bats. Careful observations from outside will help you to identify the species. You could look around, near the roost entrance, for droppings, prey remains or a dead bat. Not all baby bats that are born in a nursery roost survive, and if you find a dead one you will have an opportunity to work out which species is using the roost (see pages 46–47). You can also observe the bats in flight as they come and go, or use a bat detector to listen to their echolocation calls. Bats tend to emerge from and enter a big roost in a more or less predictable stream, which makes them easy to observe, but they may not use echolocation when they are close to the roost entrance, as they know the area so well. At dusk, you should be able to find the bats' preferred flight path from the roost to a feeding area, so you can place yourself in a good vantage point. At dawn, bats may gather and circle around the

Below: A Grey Long-eared Bat emerges from a woodpecker hole used as a summer roost.

roost before entering, providing a good opportunity for observation and a way to find new roosts. If you can get up early enough, watching bats outside a roost at dawn is an amazing experience.

Above: If you find a bat roost, it is often possible to observe the occupants entering and emerging at dusk and dawn. Here, a Greater Horseshoe Bat can be seen leaving its roost in an old barn.

If you think you have found a pipistrelle roost it is worth counting the bats as they emerge; nursery roosts of the Common Pipistrelle typically contain fewer than 100 individuals, while those of the Soprano Pipistrelle contain more. Noting the bats' time of emergence in relation to sunset might also help you to identify the species. Most species emerge on average about half an hour after sunset, though it sometimes takes a while for them all to come out, so they start earlier. Noctules tend to emerge early (at sunset), and Daubenton's Bats, Natterer's Bats and Brown Long-eared Bats emerge particularly late (about an hour after sunset). Most roosts are used by one colony of a single species of bat but occasionally two species may share.

Your local bat group will be keen to know about any bat roosts in the UK, and may wish to document changes in numbers over time, so if you do find a roost, please report it to the Bat Conservation Trust.

What's left behind

Look around in places where you suspect bats feed and you may find droppings or prey remains such as butterfly, moth or beetle wings. Outside a roost entrance is a good place to look, as the bats sometimes defecate just as they are flying out. If bats are using a church, barn, loft or porch, you may find droppings and prey remains scattered around there. If you are very lucky you might discover an outdoor perch used by a bat during feeding. Many species eat their prey entirely while on the wing, and so leave few uneaten prey remains. Bats that take bigger prey, like moths and beetles, tend to perch so that they can remove the wings and eat the softer parts of the insects. Identifying the insects is a step towards establishing the species of bat. The species most likely to leave prey remains are the long-eared bats (moths and butterflies) and the Greater Horseshoe Bat (beetles and moths).

If you find droppings, you first need to confirm that they are bat droppings rather than rodent droppings. Mice and rats mostly eat plant material such as grain; bats eat insects and other invertebrates. Insects and spiders are protected by an exoskeleton made of chitin, a hard material that is broken up into tiny, shiny particles like small grains of sand

Right: This Grey Long-eared Bat, well known for leaving prey remains, has captured an insect. The wings and remains left behind can be used to identify the prey, which in turn helps to identify the bat species.

when eaten by bats. Look closely at the droppings, using a magnifying glass, hand lens or microscope if possible, to work out whether they are made of chitin or plant fibres. Try to crumble a dropping in your fingers (wash your hands afterwards), or in a tissue if you prefer: if it is from a mouse or rat, it will feel hard and solid or it will create a smear of mashed-up plant material. If it is from a bat it will be dark, and will separate into dry, often shiny or sparkly fragments of insect chitin.

Some people claim to be able to identify the species of bat from droppings, but it is very tricky without genetic analysis. The droppings of Lesser Horseshoe Bats, the small *Myotis* species (Brandt's/Whiskered/Alcathoe Bats) and pipistrelles are usually less than 2mm (0.08in) in diameter. Those of the bigger *Myotis* species, Leisler's Bats, Barbastelles and the long-eared bats are 2–3mm (0.08– 0.11in) in diameter, and those of Greater Horseshoe Bats, Noctules and Serotines are about 3mm (0.11in) or more in diameter. The length of droppings is more variable than the diameter, and depends on the diet and the freshness of the droppings. There is not much difference in the size of the droppings of the biggest and smallest bats, but you can at least try to guess the species.

Below: Although bat droppings are relatively harmless, people may prefer not to sit on them. Sheeting, as seen here in a church in Northamptonshire, can be used to help collect and remove bat droppings.

Detecting bats

Above: This Chiffchaff looks very similar to a Willow Warbler, but the species can be identified easily from their distinctive songs.

The sounds animals produce are sometimes used to identify them. Different species of frog have varying croaks, some insects chirp in distinctive ways and, of course, bird species can be identified from their songs. Many birds produce calls as well as songs. Bird calls are brief, simple sounds used to signify alarm or hunger, while the longer and more complex songs are used to proclaim sex or species, establish ownership of a territory, find a mate, maintain a bond between mates, hold a flock together, and so on. Bird song is related to mating and identity, and some male bats also sing to attract mates in autumn.

Bats produce echolocation calls almost all the time as they fly around and, once bat detectors have been used to make the sounds audible, the calls can be used to aid species identification. However, birds are easier to identify from their song than bats are from their echolocation calls because, to fulfill their functions, the sounds birds make have evolved to differ between the species. Although echolocation calls can differ between species, echolocation has evolved to solve a functional problem that is common to all bats: it helps them find their way around, and find prey, in the dark. Echolocation calls did not evolve for communication, though some bats do listen to the calls of other bats and gain information from them. As the tasks bats use echolocation for are similar (finding food and orientation), the calls are also similar, or at least more similar than birdsong.

Despite the similarities, the bats in the British Isles can be identified to a certain extent from their echolocation calls. The best way to have a go at this is to tune a heterodyne bat detector (see page 56) to about 45kHz. At this frequency your detector will pick up any calls between about 40 and 50kHz. The echolocation calls of the pipistrelles, the *Myotis* bats, the long-eared bats and the Barbastelle will be audible as a series of fast clicks. If you hear something, tune up and down. If you hear louder, different sounds (described as 'slaps') at 45kHz, you probably have a Common Pipistrelle. If you hear louder, different sounds at 55kHz, you have a Soprano Pipistrelle. If the clicks sound similar from about 30 or 40kHz up to above 55kHz, it is likely that you have a *Myotis* bat. The long-eared bats produce very quiet calls, described as 'ticks', and you are unlikely to hear them on a detector even if you get very close to the bat. Barbastelles are also quiet, and produce sounds described as castanet-like 'smacks'.

If you hear slow, irregular series of calls that are loudest and most like 'slaps' at lower frequencies, between 18 and 40kHz, you probably have a Noctule, Leisler's Bat or Serotine. These loud calls can be detected at distances of over 100m (328ft). Noctules produce alternating calls, one at a lower frequency (e.g. 18kHz), followed by one at a higher frequency (e.g. 21kHz), sometimes described as 'chip, chop' sounds. Leisler's Bat echolocation calls are similar but higher in frequency.

If you hear continuous warbling sounds, rather than clicks, you may be picking up a horseshoe bat. The two species are the only bats in the British Isles that can be identified with certainty using a simple heterodyne bat detector. The Greater Horseshoe Bat's calls are loudest when you tune to 83kHz; the Lesser Horseshoe Bat's peak at 112kHz. These sounds are quite loud but are directional and dissipate rapidly in air, and so are hard to detect. You have to be right in front of the bat to hear them.

When you are out with a bat detector, you may pick up confusing ultrasounds produced by keys jangling, crickets chirping, shiny fabric sliding against itself (for example, the sleeves of your coat as you walk), and so on. You may also hear social calls produced by pipistrelles for communication at around 15kHz, or special sounds made by feeding bats as they hone in on insects. These sounds are called feeding buzzes and sound like a person blowing a raspberry.

If you are interested in learning more about identifying echolocation bats from their calls, it is a good idea to go on a bat walk with an expert or buy a book specifically about this subject.

Below: A tunable heterodyne bat detector like this Batbox III can make ultrasonic calls audible to humans.

IDENTIFYING BATS

Bat detectors

Bat detectors transform the ultrasonic echolocation calls produced by bats that are too high-pitched for us to hear, and make them audible. Like other electronic equipment, bat detectors are getting smaller. In the 1950s, the equipment needed to transform and record echolocation calls just fitted into a pickup truck. Now it can be carried in your pocket. You can even buy a tiny plug-in device for your smartphone or tablet that allows you to hear, visualise and record ultrasounds, and weighs less than a Noctule!

Thankfully, bats are not disturbed by humans walking around pointing bat detectors at them. A bat detector opens up a whole new world to the user. Bats seem to be everywhere as their sounds tell you where to look to see them, and help you follow their flight paths with your eyes. The echolocation calls are often beautifully musical, while calls used for communication, or social calls, which can also be heard on a detector, sound raucous and harsh. As different species of bat use different sounds for echolocation, it is sometimes possible to work out which species of bat is flying overhead from the sound heard on the detector (see pages 54–55).

The best type of bat detector for casual observation and identification by a beginner is the heterodyne one, which makes the ultrasounds audible by mixing them with other sounds that the detector generates.

Heterodyne detectors can be tuned to pick up sounds of different frequencies. This has the disadvantage that they do not pick up any bats calling at frequencies outside the range they are tuned to, and the advantage that they allow the user to work out the frequency at which a bat is calling. There are also heterodyne detectors that tune themselves automatically, based on the calls they pick up, and report the loudest frequency to the user via a screen. The frequency provides important information for identification.

Frequency division detectors work by dividing the frequency of sounds by a given number, so that high-frequency ultrasonic sounds become

Above: A bat detector makes it possible to hear the ultrasonic echolocation calls that bats make. Once bats can be heard, they are easier to see. This detector, the QMC Mini, was one of the first affordable commercial models; it was available from the 1970s.

audible. A wide range of frequencies are made audible, but it is impossible to hear what frequency the bat is calling at. So, no bats are missed, but they cannot be identified unless recordings are made and analysed later, and in recordings much of the fine detail of each call is lost. Frequency division detectors are often left to record in the field for several nights or weeks at a time. This produces valuable standardised information about the level of bat activity in one location, but a person walking around can often find more bats by responding to sightings or sounds. Some detectors offer both heterodyne and frequency division functions, allowing all frequencies to be scanned so that as many recordings as possible are collected.

Time-expansion detectors work by effectively slowing down the calls, so that they become longer in duration and lower in frequency. This was originally achieved by recording on large reel-to-reel tape decks and playing back the recordings at a slower speed. The method was replaced by a digital one, in which a short recording was made and then played back more slowly, resulting in recordings that were not continuous but in which all the detail of the calls was preserved for analysis.

Now that data storage space is much more readily available, continuous recordings can be made in real time using a method called direct sampling. The calls can be played back slowly, so that, as with time expansion, the entire structure of the calls is preserved. Direct sampling of ultrasound is ideal for detailed sound analysis and for identifying bats, and can be used in the field with a smartphone, tablet or laptop.

Above: This detector can be left out overnight to record echolocation calls, which can be downloaded, analysed and identified.

Bats Around the World

Though all bats share certain features – furry bodies, wings and a nocturnal lifestyle – as a group, they are very diverse. Some bats roost in trees or caves, others in burrows or even in spiders' webs. Some bats live alone or in small groups, while others live in colonies that may consist of millions of individuals. Bats can also occupy many different habitats, including rainforests, farmlands, wetlands and deserts. In keeping with the variety of directions in which evolution has taken them, bats are various colours, shapes and sizes and, in this chapter, I'm going to describe a few of the bats I find most fascinating.

Bats with special grip

Most bats use their claws to grip rough surfaces when they are resting; however, there are seven species of insectivorous bat that have evolved suction pads. Unusually for bats, these species roost with their heads pointing up, and use sweat and suction to stick themselves to smooth leaves. Two species of sucker-footed bat are found in Madagascar – the Madagascar Sucker-footed Bat (*Myzopoda aurita*), which is also known as the Eastern Sucker-footed Bat, and the Western Sucker-footed Bat (*Myzopoda schliemanni*). They hold on to the smooth insides of the rolled-up leaves of Traveller's Trees (*Ravenala madagascariensis*) not by creating suction but by wet adhesion: a liquid similar to sweat is produced on the pads on their wrists and ankles. This method means that they can only roost with their heads up.

Opposite: A colony of fruit-eating Honduran White Bats in the tent they have constructed by biting a leaf in strategic places.

Below: The Traveller's Tree is a favourite roosting place of sucker-footed bats.

Above: At the Smithsonian Tropical Research Station, Barro Colorado Island, Panama, researchers placed this Spix's Disk-winged Bat inside a glass cylinder so that they could examine the suction pads on its wings and hind feet.

Spix's Disk-winged Bat (*Thyroptera tricolor*), found in Central and South America, also typically roosts head-up, though it can roost head-down as well. It has disks on the base of its thumb and under its heel that are concave (curving inwards, forming hollows). This bat uses a special muscle to change the shape of the disk to create or release the suction. The disks are kept moist by the action of sweat glands, and the bat uses them to cling to the smooth inside surfaces of rolled-up leaves, often of *Heliconia* (wild plantain). These leafy, tube-shaped roosts are inhabited by several bats at a time, roosting one above the other, but each leaf soon unfurls and so is often used for only one day.

Right: The Spix's Disk-winged Bat prefers to roost inside a rolled-up leaf, where its special adaptations allow it to stick to the leaf surface.

Fishing bats

The Greater Bulldog Bat (*Noctilio leporinus*) is a large piscivore, which means it eats mostly fish. Hunting over open water in Central and South America and the Caribbean, it finds the fish by detecting the echoes that tiny ripples create on the water surface. Greater Bulldog Bats have huge, strong hind claws to grasp the fish, large tail membranes, which they use to trawl through the water, and cheek pouches to store food. They vary in colour, from pale grey to dark orange, with dark grey or brown wings. Their roosts have a distinctly musty, fishy smell.

Below: The Greater Bulldog Bat fishes in lakes and the sea, using its large hind feet to trawl through the water to catch prey.

Vampires

Strictly speaking, the Common Vampire Bat (*Desmodus rotundus*) is not a bloodsucking creature, but a blood-licker. It uses echolocation and its sense of smell to find prey, often landing on the ground nearby and approaching the target animal by crawling. For a bat, crawling means walking on its hind feet and thumbs, with wings folded; vampire bats are very good at crawling, and can also bound or jump. In common with fruit bats that crawl around a lot on vegetation, vampire bats have only very tiny tail membranes. A big flap of skin would get in the way.

The Common Vampire Bat, the Hairy-legged Vampire Bat (*Diphylla ecaudata*) and the White-winged Vampire Bat (*Diaemus youngi*) are the only mammals that feed exclusively on blood, and the only mammals known to have a heat-sensing organ, but they have a very limited sense of taste. Once the bat has located a suitable victim, perhaps a sleeping cow, horse or pig, it uses the heat-sensitive part of its nose to find a warm area of skin that is rich in blood vessels. Once it has selected a suitable place to feed, the bat prepares the chosen spot by licking the skin and by biting off and spitting out the fur. It then makes a small cut with its sharp incisors and laps up the blood using its specially adapted tongue, which has grooves that work like a drinking straw. As a vampire bat closes its mouth, its teeth hone each other, so they are always razor-sharp. Vampire bat saliva contains anticoagulants that prevent the victim's blood from clotting, sometimes also after they have finished feeding, so the wounds they inflict may bleed for a while. When a vampire bat returns to a feeding place on an animal that it has used before, as it often does, it can start to feed much more quickly upon simply reopening the wound.

Common Vampire Bats can go a few days without food and can

Below: Common Vampire Bats that roost together get to know each other well, and even help their roost-mates if food is scarce.

drink almost their own body weight (35g or 1.2oz) in blood (25ml or 0.9fl oz). In colonies, Common Vampire Bats are very sociable; they get to know each other and can recognise other individuals, which allows them to regurgitate blood to feed their friends, neighbours or relatives who have failed to find food on a given night. Each bat is most likely to supply other bats that have fed it in the past, hence the need for individual recognition. Females also feed their babies in this way, and adults feed bats that have been orphaned and are still too young to find their own blood meals.

Though Common Vampire Bats are more likely to feed on farm animals than on humans, many people living or travelling in Mexico and other countries in Central and South America, where these bats are found, worry about being bitten. Common Vampire Bats can transmit diseases, including rabies, and many bats have been killed to try to reduce the spread. In general, though, their victims sacrifice only a small amount of blood and suffer no ill effects – and they don't develop a taste for blood themselves!

Above: Common Vampire Bats often feed on the blood of domestic animals, in this case a goat.

Nectar-eating bats

Plants don't move around so in order to generate seeds to make the next generation, most flowers need to be pollinated by species that can move from flower to flower: moths, butterflies, other insects, birds or bats. Pallas's Long-tongued Bat (*Glossophaga soricina*), found in Central and South America and sometimes kept in zoos, feeds on nectar and pollen. It weighs about 10g (0.4oz). When feeding, it flies from flower to flower and its fur becomes covered with pollen. Hummingbirds feed in a similar way: hovering in front of each flower, they use their beaks and tongues to collect nectar, while at the same time transferring pollen between the flowers. In common with other nectarivorous bats, Pallas's Long-tongued Bat can hover, and it has a long nose, a hugely long tongue covered with special bristles to mop up nectar and few teeth, as it does not need to chew. This bat has the fastest metabolism ever recorded in a mammal, which means that it converts its food to energy very quickly. Consequently, it needs to feed frequently and speedily, and cannot survive for long without sugary, energy-rich nectar.

Below: A Pallas's Long-tongued Bat in Panama, feeding on nectar from a *Pseudobombax* flower.

The Tube-lipped Tail-less Bat (*Anoura fistulata*) also feeds on nectar. It lives in cloud forests in the Ecuadorean Andes and was not described by scientists until 2005. It has a very long nectar-licking tongue, the longest of any mammal in relation to its body size: 8.5cm (3.3in) or 1.5 times the length of its body. One species of plant, *Centropogon nigricans*, has evolved to produce nectar at the end of long tubes in its flowers so only this species of bat can pollinate it, and the bat has special adaptations in its ribcage to allow the tongue to be stored when it is not in use.

Tent-makers

The rare Honduran White Bat (*Ectophylla alba*) has white or grey fur and its ears and nose-leaf are bright yellow. Its wings are mostly black. It roosts under the leaves of *Heliconia* or other plants, which it modifies to make a tent. It chews through the side veins of each leaf parallel to the central vein so that a leaf folds over on each side and creates an extra warm and sheltered roosting place for each small colony – usually one male and his harem of around 3–7 females. Sunlight filtering through the leaf gives the bats' white fur a greenish appearance, providing good camouflage. These beautiful little bats weigh about 5–6g (0.2oz) and the IUCN classes them as Near Threatened. Numbers have declined due to habitat loss and urbanisation: the growing human population is a major threat to this species. The Honduran White Bat is found in Central America – from Honduras to western Panama – in lowland forest areas. It eats fruit, including wild figs; the tiny seeds pass through its body and are deposited with a small amount of fertiliser so that the bats contribute to the dispersal of these plants.

Below: Honduran White Bats roost close together in the tent-like shelter they create. The inverted shape of the leaf provides protection from the sun, rain and predators.

Record-breakers

Mexican Free-tailed Bats (*Tadarida brasiliensis*), found in North, Central and South America, use Bracken Cave in Texas, USA, as a summer nursery roost for breeding, along with many other caves and buildings. At Bracken Cave, they form a gathering of 5–20 million female bats; the number approximately doubles in June when the bats give birth. This colony is believed to be the most numerous group of mammals in the world; in some years it is more numerous even than the human population in the biggest cities. Biologists working here observed that when giving birth, each mother bat clings on to the roof of the cave with both thumbs and one or both hind feet, then remains attached to her naked baby for an hour or so while she grooms and feeds it for the first time. Mother and baby need to get to know each other's scent and sound well so that they can find each other among their millions of roost-mates. The babies tend to roost together, away from the adults, and each mother goes out to feed every night. On her return, it often takes a few minutes for the mother to find her baby using first the sound of its call to locate it, then its smell to confirm its identity.

Below: This insectivorous Mexican Free-tailed Bat in Texas, USA, illustrates its name by showing its tail.

At dusk, the cloud of bats emerging from Bracken Cave is so vast that airport and weather radar systems can detect it. Each brownish adult bat is small, weighing around 12g (0.4oz) with a wingspan of about 28cm (11in), but together, 20 million Mexican Free-tailed Bats are estimated to eat over 200 metric tons of insects every night!

The Mexican Free-tailed Bat is one of the fastest flying animals on earth: it can reach horizontal speeds of up to 160km/h (100mph) in level flight, though it is possible that the bats make use of tailwinds to achieve the fastest speeds. It also flies at the highest altitude recorded for any bat: up to 3,300m (11,000ft). To avoid the cold weather in the USA, many bats of this species migrate

from the USA to Mexico and other countries in Central America and stay active all year round, while other species hibernate.

The Mexican Free-tailed Bat and its larger relative, the European Free-tailed Bat (*Tadarida teniotis*), have very long, narrow wings that allow them to fly fast. European Free-tailed Bats fly at speeds of well over 50km/h (31mph), at high altitudes, while feeding on clouds of high-flying insects. Their mouse-like tails stick out beyond their tail membranes. The European Free-tailed Bat is found around the Mediterranean Sea and can be recognised by its distinctive high flight. It weighs 28–57g (1–2oz) and has a wingspan of up to 45cm (18in). Its long echolocation calls are audible to many people, as they range in frequency from 10–18kHz (see page 17).

Above: Mexican Free-tailed Bats emerging from Bracken Cave, Texas, USA, where the summer colony is believed to be the biggest gathering of mammals in the world. There are so many bats here that they have to start emerging before dusk.

Croak and you're history

The Fringe-lipped Bat (*Trachops cirrhosus*) weighs about 32g (1.1oz) and has reddish-brown fur. It inhabits forests in Central and South America, from southern Mexico to Bolivia and southern Brazil, and has large ears and a long nose-leaf. It is an omnivore, taking what it can glean from the forest floor, and it is particularly keen on frogs. When feeding on frogs, the Fringe-lipped Bat listens carefully to the mating calls the male frogs produce, using them to locate their prey and to identify edible frogs while avoiding poisonous or distasteful ones. It can even use the call to assess the size of a frog so that it can bypass those that are too big to handle. In response to this hunting method, frogs have evolved to use short calls that are difficult to locate.

Below: The Fringe-lipped Bat listens intently for the mating calls that frogs produce; it uses this sound to locate its prey.

Foraging in the forest

The New Zealand Lesser Short-tailed Bat (*Mystacina tuberculata*), a declining species that the IUCN classes as Vulnerable and that is found only in New Zealand, eats everything it can find – mainly insects (beetles, moths and flies) but also fruit, nectar and pollen. It roosts in old, hollow trees and, unusually for a bat, forages for insects and plant material while rooting around in and under the leaf litter on the forest floor, as well as while flying. It is an agile walker, with strong hind legs and feet; it folds its wings tightly when on the ground so that it can use its forearms to walk. It uses echolocation when flying, but this is no good for locating prey under leaf litter, so it sniffs out its prey and listens for the low-frequency rustling sounds that moving insects produce. It weighs 10–22g (0.4–0.8oz) and has distinctive tube-like nostrils. It is one of only two endemic land mammals in New Zealand – both are bats, and they are the only two species of bat currently found in New Zealand. During their evolution, there were no competing insectivorous mammals, and there were certainly no predators on the forest floor, so the bats could forage there with little risk. However, the species is now threatened by the loss of the mature forest it needs and by introduced predators such as rats, domestic cats and Stoats (*Mustela erminea*).

Above: A Lesser Short-tailed Bat forages on the forest floor in New Zealand. The tube-like nostrils are distinctive, and the wings are tightly folded to enable it to walk.

Lifestyle and Behaviour

Individual bats live for many years, and each year follows a pattern. There is an annual cycle of mating, hibernating, giving birth and feeding. Bats are sociable creatures. Seeking safety in numbers, female bats in Europe spend the summer in busy nursery colonies, where they can exchange information, look after their babies and huddle together or spread out as appropriate, in order to maintain an ideal body temperature. At the end of summer, nursery colonies break up and bats start looking for mates, and then hibernation sites. The annual cycle continues.

Tiny but long-lived

Bats are unusual: even though they are small animals, they breed slowly and live for a long time – about ten times as long as shrews or mice of a similar size. Greater Horseshoe Bats can live into their thirties, and the maximum documented lifespan for most other species from the British Isles is about 20 years. Outside the British Isles, small bats have been documented to live for twice as long. Bats may have evolved to live long lives so that they can produce several young in their lifetime. Carrying extra weight is hard for flying animals, especially small ones, so being pregnant with a large litter would be particularly demanding and risky. Birds have solved this problem by laying eggs instead of carrying young. Most bats, certainly those living in colder climates, typically have only one baby annually. Though the first few months of life and the first winter are critical times when many young bats die, if they do survive, their chances of having a long life are good. Adult survival is high because predators find bats hard to catch. In addition, hibernation allows bats to slow

Opposite: Bechstein's Bats have been recorded as living up to 21 years, which is a typical lifespan for all the bat species living in the British Isles.

Below: This Noctule has been caught and fitted with a ring so that biologists can collect information if it is caught again. Ringing studies have revealed that bats are extremely long-lived for their size.

Above: For female bats like this Northern Myotis, flying while heavily pregnant must be hard work.

down when food is scarce, saving energy, so that they can live for longer. A ringed Brandt's Bat from Siberia lived for at least 43 years; no doubt it had hibernated for much of its life and had found plenty of insects in the short summers. However, even bat species that don't hibernate live relatively long lives.

The gestation period (pregnancy) is longer in bats than in other small mammals: even in very small bats it can be a couple of months. In comparison, the gestation period in mice is 20 days; the average litter size is 6–8. Young bats grow up slowly, stay with their mothers for a long time and do not usually mate until they are at least nine months to a year old. Female bats often continue to have a baby a year until they die.

Because individual bats live relatively long lives, bat numbers in most areas are more stable than numbers of other small mammals. Unfortunately, their lifestyle also means that bat populations cannot recover quickly from declines. Bats simply cannot reproduce fast enough to increase their numbers speedily if, for whatever reason, many of their community are killed. This means that bat conservation is a challenge.

What do bats die of?

Worldwide, many bats are killed by humans, including vampire bats because they are blamed for spreading rabies in Central and South America, and fruit bats because they are blamed for eating crops in Asia and Australia. Fruit bats are killed and eaten by humans in Africa and Asia. In the last 15–20 years, increasing numbers of bats have died by colliding with wind turbines. Bats sometimes die when they fly into power cables, vehicles, barbed wire fences or spiky plants, and bats caught in heavy rainfall may become waterlogged and die.

Individual bats may die of starvation for many reasons: if poor weather conditions or habitat changes result in declines in their insect prey; if energy demands during hibernation are too high because of disturbance by humans; if they fail to find a suitable roost; if their roost site is destroyed; or if they run out of energy or get blown off course during migration. Bats are also taken by predators or poisoned by pesticides and by chemicals used for timber treatment in the buildings in which they roost, though some of the most toxic timber treatment chemicals have been withdrawn from sale.

Like other mammals, bats can carry various viruses, including a type of rabies, but these viruses often do not seem to cause sickness or death, so that researchers are very interested in the immune systems of bats. A disease called white-nose syndrome, caused by a fungal infection, was first diagnosed in bats in 2006 and has since resulted

Left: This hibernating Little Brown Bat in Vermont, USA, has white-nose syndrome and will not survive the winter. The fungus causing the disease has evolved to exist in dark caves, and is destroyed by ultraviolet light.

Above: Parasites, like this wingless bat-fly (left), have evolved to feed specifically on bats. Ticks also feed on bats; in this case, a Greater Mouse-eared Bat (right).

Below: A Whiskered Bat that has been caught and injured by a cat is examined at North Devon Bat Care, a bat rescue centre in the UK. Small holes in the wings can heal quickly, but there may be other, more serious, injuries.

in the death of millions of hibernating bats in the USA and Canada. The fungus, but not the disease, has been found in bats in Europe. Scientists believe that the fungus originated in Europe, where the bats are naturally resistant to it. Bats are also fed upon by various creepy-looking parasites, including fleas, bat-flies, bat-bugs, ticks and mites. In most cases these do not cause death, but some bats die of diseases carried by parasites.

It is often hard to pinpoint one reason for the death of an individual: a young bat may not find enough food, then become susceptible to disease and parasites, and then, in its weakened state, it may be caught and killed by a cat.

A bat's year

For a big chunk of the year, generally from around mid-November to mid-April, most bats in Europe are hibernating. At this time it is often cold and wet, and few insects are available. Hibernating bats arouse at least briefly every few weeks during the winter; some arouse to feed all through the winter on nights when it is not too cold. Insectivorous bats lose about one-quarter of their body weight over the winter during hibernation. From around March, bats are active on warm nights, and some mate at this time too. Common Pipistrelles are among the first species to leave the hibernation roosts in spring; *Myotis* bats tend to leave later, at the end of April.

In around mid-April or May in Europe female bats ovulate and become pregnant; the gestation period is 6–10 weeks depending on the species. Most bats will have mated in late summer or early autumn, and females store live sperm in their bodies until they need it. This means that as soon as the females emerge from hibernation in spring, they can become pregnant without the time-consuming hassle of finding a mate. Instead, they can concentrate on feeding to regain the weight they lost during hibernation.

Below: In winter, colonies of Greater Horseshoe Bats use caves as hibernation sites.

Above: A nursery colony of Greater Mouse-eared Bats (now extinct in the British Isles), consisting of females with their young, in the roof of a church in Germany.

From about May, females group together in nursery colonies. Males often roost separately in small groups or alone. Baby bats, known as pups, are born in late June, and their mothers feed them milk for several weeks. They start to fly in July or August, usually 3–4 weeks after birth, and gradually become independent as milk production slows down and eventually stops. Weaning onto a diet without milk is complete around 6–9 weeks after birth.

In August or September, the nursery colonies break up, and in September and October, the priority for bats is to fatten themselves up ready for hibernation. They start to slow down their metabolism by using torpor during the day, to conserve energy. Mating is also on the agenda at this time of year: males start swarming outside hibernation sites or displaying at roosts used specifically for mating, and mating takes place at swarming sites, mating roosts or while bats are migrating.

In October and November, bats throughout Europe start to find and use their hibernation roosts.

Finding a mate

Since female bats can store live sperm within their bodies until it is time for them to become pregnant in spring, mating can take place at the end of summer, or in autumn, winter or spring, though autumn is the main mating season in Europe. Females assess potential mates carefully and select the strongest, healthiest and most attractive male to become the father of their offspring. Males, if they are successful, can mate with several females each year and father several young, so they are naturally less selective. In autumn, male long-eared bats and some *Myotis* bats seek out females in maternity colonies and presumably mate there. Other males advertise themselves to females by swarming, songflighting or calling from mating roosts.

In swarming species, such as most *Myotis* bats, males gather in places where they can expect to find females, often near the entrances to popular hibernation sites such as big caves. In some places, many bats of several species fly around together, creating a large and spectacular swarm in the middle of the night – bats travel a long way to reach these special swarming places. Many female bats visit the swarming site, perhaps on their way to investigate the possible hibernation site, and can inspect, assess and select the males as they do so.

Below: Bechstein's Bats often mate in autumn. Here the male, whose penis is visible, is chasing the female.

Above: Male pipistrelles often conduct songflights near street lamps. Students from Bath University, UK, have placed an automatic bat detector on this lamp in order to monitor bat activity.

Male bats of other species advertise themselves by conducting songflights while producing special calls in order to attract females. If you go out with a bat detector in September and find a street lamp in a village or small town, you may find that a male pipistrelle is using it as a place to perform his songflight. You may see him circling around the light, making loud calls at about 15kHz every so often; some people with good hearing can hear the calls without a bat detector. In between these songflight calls you may hear echolocation calls at much higher frequencies, but for this, you will need a bat detector. Females feeding around the street lamps can assess their options and select a male; they then roost nearby with him for a few days.

In other species, such as Noctules and Nathusius' Pipistrelles, the males seek out and occupy special mating roosts, little love-dens from which they call to the females. Mating roosts are often small, cosy tree-holes, perhaps just big enough for two fairly intimate bats, or for a male with a few females. On autumn evenings, the male sits near his roost, calling to make himself look and sound as attractive as possible, until a female comes to join him.

Migration and hibernation

All European bat species feed on insects and other invertebrates that are hard to find in the colder months, and so all of them hibernate. Hibernation sites need to be humid (otherwise bats would dry out), have constant temperatures (between 1–12°C, or 34–54°F), and be relatively undisturbed and safe from predators. So hibernation sites are scarce, especially for those bats that choose underground sites. Some bats travel a long way to a suitable site, and some migrate in a predictable way every autumn, returning to their summer feeding areas in spring.

Bats use echolocation for orientation but not for long-distance navigation (for example, during migration). Instead, they navigate like birds: migrating bats can detect the Earth's magnetic field and use it during long-distance flights.

Horseshoe bats, long-eared bats, Bechstein's Bats and Natterer's Bats do not migrate long distances as far as we know. All of the other species migrate to their hibernation sites in at least some parts of their ranges, though not necessarily in the British Isles. For example, it is believed that Noctules are sedentary in the British Isles, while in

Below: Suitable hibernation sites are rare. This Lesser Horseshoe Bat, a non-migratory species, is hibernating among stalactites in a cave in Sardinia, Italy.

Above: One of the largest examples of bat migrations worldwide is that of Straw-coloured Fruit Bats in Zambia. Unlike European bat species that migrate to their hibernation sites, these tropical bats migrate to locations where food is more abundant, contributing to seed dispersal along the way.

continental Europe they breed in the north or north-east but mate and hibernate in the south. Leisler's Bats and Nathusius' Pipistrelles are also long-distance migrants in continental Europe at least. They regularly travel 1,000km (621 miles) twice a year, moving towards the south-west in autumn to hibernation sites, and back to the north-east in spring.

Migration is not always to and from hibernation sites. Tropical and subtropical bats migrate to places with more abundant food, and some North American bats migrate to warmer areas in the south and therefore avoid hibernation.

Family life

From around May, female bats in Europe start to gather together in nursery roosts. These are busy places, containing from tens to hundreds of female bats. Soprano Pipistrelle nursery roosts may contain over 1,000 bats. The number approximately doubles when the babies are born. Depending on the species, bats in Europe roost in tree holes, bat boxes, cavities behind bark, buildings and holes in structures such as bridges. Common and Soprano Pipistrelles tend to use buildings, whereas Daubenton's Bats favour bridges. Tree-roosting bats tend to move a lot, shifting their roost every few days, perhaps to avoid the build-up of parasites. Bats in buildings are more likely to stay put for the pup-rearing period.

Most female bats give birth to one young in June, just in time for the bumper insect months of June, July and August. Some species, notably Noctules, Leisler's bats and pipistrelles, occasionally have twins, particularly in continental Europe. Singleton pups weigh 20–30 per cent of the mother's weight, which is extremely high (for comparison, an average newborn baby human weighs around 5 per cent of its mother's weight). Pups are born blind and naked, and spend a lot of time clinging to their mothers until their eyes open and they start to grow fur; this happens within a few days. They are normally left in groups or crèches in the roost when their mothers feed, and are only carried around when the mothers move to a different roost.

Left: Young Common Pipistrelles, seen here, are fed solely on their mother's milk for the first few weeks of their lives. They gradually learn to fly and hunt for themselves as milk production declines.

Above: An adult female Common Pipistrelle with her pup in a nursery roost in Germany.

Opposite: This female Lesser Horseshoe Bat is hanging from a roof with her pup.

Mothers and pups must be able to recognise each other, as each female suckles only her own young, and the mother needs to find her pup in the crèche when she returns to the roost a few times each night after feeding. Within a few hours of birth, pups produce special isolation calls, which their mothers recognise, and mothers and pups also use their keen sense of smell to identify each other. Once fully grown, female pups tend to return to the roosts they were born in to give birth themselves, so they may well meet their mothers, grandmothers, sisters and aunties.

As well as resting during the day, bats in nursery roosts socialise, squabble, mate, groom themselves and each other, and suckle their young. Bats in colonies communicate with one another, and hungry bats follow others that have been successful in hunting. Dads don't get involved in family life; in fact, male bats often avoid nursery roosts.

Several species of bat move around all summer, so changing roosts is not a big issue. But in most cases, the colony remains together and moves to a new place en masse. At the end of the summer, nursery roosts break up completely and mothers, youngsters and roost-mates presumably go their separate ways. Some independent young bats do remain loosely associated with their mothers over several years, and may learn about good feeding and roosting sites by foraging and roosting with them at least some of the time.

Food for Bats and Bats as Food

Worldwide, bats take a wide range of high-energy foods; they need to fuel their rapid and energetically demanding flight without weighing themselves down by eating bulky food that is hard to digest. Bats have evolved various methods to find, and capture, their food. Fruit, nectar, pollen, blood and other animals are all energy-rich enough to be on bats' menus. Most bat species have evolved to make use of insects and other invertebrates: small and energy-rich but challenging to capture. Bats themselves fall prey to various predators and have developed ways to avoid being eaten.

In the tropics you'll find bats eating the most exciting foods, as well as the most thrilling bat-eating animals. Here, bats feed on fruit, flowers, fish, blood, frogs and insects… but it is also here that bigger bats, spiders, birds and snakes eat bats. In Europe, where most bats feed on insects and have few natural enemies, there are still birds that eat bats, while bats of another species get their own back by eating birds; there are also fish-eating bats!

Opposite: This Brown Long-eared Bat has caught a yellow underwing moth and taken it to a feeding perch, where it will consume the body and discard the wings.

Left: Pallid Bats like this one in Mexico are mostly insectivorous, feeding on crickets and scorpions, but they also eat nectar in spring. This one is covered in the pollen of a cactus that has a cone-shaped flower suitable for pollination by bats.

Foraging strategies

The flight patterns and echolocation calls that bats use have evolved together with their diet. To feed on different foods in varying situations, bats use five main foraging strategies, known as fast hawking, slow hawking, trawling, gleaning and perch-hunting. These strategies can be combined so that bats use more than one while hunting. However, flight ability is related to wing shape, body weight and tail membrane shape so, for example, only bats with relatively large, rounded wing tips can hover and glean, and these bats cannot fly fast. Bats with long, narrow wings and pointed wing tips must keep moving relatively fast or risk dropping out of the sky; they are experts at fast flight but generally avoid dense vegetation, as their flight is not very manoeuvrable. Echolocation calls have evolved to fit in with wing shape and flight style, so gleaning bats tend to have quiet calls suitable for close targets, while fast-hawking bats have loud calls that can travel to distant targets.

Fast-hawking bats fly quickly in pursuit of flying insects, birds or bats, relying on speed and agility to catch their prey. They use loud, long-ranging echolocation calls, often of relatively low frequency. Slow-hawking bats hunt while flying more slowly and with greater manoeuvrability, often

Below: This Brown Long-eared Bat is hunting by slow hawking, as it flies close to the vegetation. It may use gleaning to scoop up its prey.

closer to vegetation, detecting prey at short range. Trawling bats, such as Daubenton's Bats and the fishing bats, pick up prey from the surface of water with their feet or tail membrane. They often have big feet and claws. Gleaning bats take resting or non-flying prey items from leaves or from the ground, and many are capable of hovering, which is also useful for collecting nectar from flowers; nectarivorous bats are also gleaners. Some gleaners land on the ground or on plants while hunting. They often use brief, quiet echolocation calls that are suitable for detecting close targets and are unlikely to be heard by prey. Perch-hunting bats spend much of the time hanging in a suitable place, using echolocation or vision to seek prey. By using long echolocation calls of mainly a single frequency, these bats can perceive echoes from fluttering prey. When they detect a suitable prey item, they leave their perch to catch it, then return to the perch to eat it.

Bats can catch prey in their mouths. Some use the claws on their feet, their tail membranes or their wings to take big prey items. Any items of prey not caught in the mouth are transferred to the mouth for eating; this is often done in flight, though perch-hunting bats may return to their vantage point before tackling their kill.

Above: A Daubenton's Bat hunting by trawling. It drags its large, hairy feet and its tail membrane through the water to capture an insect on the surface.

Food for bats

Bats like their food to be nutritious and high in calories. Some bats are highly specialist feeders, others are flexible and can make use of whatever food is most readily available. Bat species can be insectivorous (eating insects or other invertebrates), frugivorous (fruit-eating), nectarivorous (feeding on nectar and pollen), piscivorous (fish-eating), carnivorous (eating other bats, frogs or other small animals) or sanguivorous (blood-feeding). Most bats are insectivorous. However, lots of species eat plants and many are omnivores, which means they eat a wide range of foods – in most cases they eat insects, fruit and nectar. In fact, it is extremely likely that the mainly vegetarian bats consume any insects that happen to be living on their food, so their diet is not as specialist as it may seem.

Around 70 per cent of the almost 1,400 bat species worldwide are insectivorous, and only insectivorous bats are found in the British Isles. In summer, each insectivorous bat eats a quarter to a third of its body weight in insects and/or other invertebrates each night. For obvious reasons,

Right: This captive fruit bat is eating a slice of fruit that a butterfly was also feeding on. It is easy to see how the frugivorous bat might eat the insect that has become trapped under its wing.

bats often eat insects that fly at night, such as moths and midges and other small flies; beetles are also popular prey. However, gleaning bats can take insects and other invertebrates that do not fly, as well as insects that only fly during the day. For example, Natterer's Bats eat a lot of diurnal house flies, which they scoop up from their nightly resting places by using the bristles on the edge of their tail membrane. Several of the *Myotis* bats eat spiders and other non-flying prey, caught by gleaning.

Moths are mostly larger than midges and are nocturnal, so they are perfect bat food. But not all bats can feed on moths, and for at least 50 million years, bats and moths have been involved in an 'evolutionary arms-race'. Some moths developed the ability to hear the loud echolocation calls that bats produce, which gave them an opportunity to

Above: The European Free-tailed Bat (left) is a moth specialist. Invertebrates taken by bats include the Buff Ermine (top), a hearing moth that Barbastelles eat; midges (middle), common prey for smaller bats; and Autumn Spiders (bottom), taken by some *Myotis* bats.

avoid being caught and eaten by their bat predators. If they hear ultrasound, these moths suddenly change direction or drop out of the sky, which makes them hard to catch. In an evolutionary counter-move, some bats developed echolocation calls that moths cannot hear. Some moths can produce ultrasound themselves, to confuse or startle bats by 'jamming' their echolocation, and some stealthy moths developed wings that can absorb ultrasound, so that they are inconspicuous to bats. Despite the ongoing evolutionary arms-race, there are specialist moth-eating bats, such as the Barbastelle, the long-eared bats and the fast hawking free-tailed bats.

The frugivorous and nectarivorous bats, and the omnivorous bats that feed on flowers and fruit, are important pollinators and seed dispersers. Flying from flower to flower when feeding on nectar and pollen, nectarivores get covered in golden pollen grains, which are transferred to other flowers and fertilise them. Flowers have evolved forms, scents and patterns that attract pollinators and facilitate pollen transfer. Many flowers 'fit' a certain pollinator, as pollen transfer is more effective if only one or a few species of pollinator is involved. Over 500 species of plant have evolved to rely on bats for pollination.

Bat-pollinated flowers, unsurprisingly, tend to open at night. They are often pale in colour, so that they stand out

Below: A Lesser Short-nosed Fruit Bat carrying a fig away from the tree to consume elsewhere. The bat will disperse the seeds of the fruit.

Above: Jamaican Fruit Bats eat nectar as well as fruit. The members of this group have returned to their roost covered in grains of pollen.

in the darkness, though some are drab and dull. Some bat-pollinated flowers have evolved to produce clear echoes, so that they act as targets to echolocating bats. The flowers are big and robust enough to be accessible to bats; often they form on branches or trunks rather than on thin twigs among leaves, and they have a musty smell. Some are cone-shaped and big enough for a hovering bat to fit its nose, or even its head and shoulders, in; others are bottle-brush shaped, allowing bats to become covered in pollen while brushing past the pollen-producing stamens to reach the nectar.

Frugivorous bats often move the seeds of the fruit they eat from one place to another, so playing a vital role in plant dispersal; seeds may even germinate better after passing through the digestive tract of a bat. Flying foxes eat the fruit of over 600 plant species. Agave, kapok, fig, eucalyptus, durian, mango, clove, banana, guava, avocado, breadfruit, ebony, mahogany and cashew trees all rely on bats to allow them to colonise new areas.

Seed dispersal takes place when bats either drop uneaten seeds, or eat seeds but don't digest them. Some frugivorous bats also eat leaves by chewing them, swallowing the protein-rich juice and spitting out the indigestible fibrous parts.

Right: The Greater Noctule is one of a small number of bat species that hunts birds on the wing; its long, narrow wings have evolved for open-air hunting.

Below: The Greater Noctule eats Blue Tits like this one, but closely related Great Tits eat hibernating bats!

Carnivorous bats – there are around 15 species worldwide – tend to be relatively large, to enable them to tackle prey such as other bats, rodents, frogs and lizards. However, the Common Big-eared Bat (*Micronycteris microtis*) of Central and South America weighs only 5–7g (0.2oz) and eats tiny lizards as well as insects. Carnivorous bats ambush prey, catch them in flight, or grab them from the ground or from other surfaces, and usually eat the whole animal, including bones and feathers, fur or scales. The Greater Noctule (*Nyctalus lasiopterus*) looks like our familiar Noctule but is a lot bigger (it weighs 35–60g or 1.2–2.1oz). It does eat insects but it also eats over 30 species of small bird, including Robins (*Erithacus rubecula*) and Blue Tits (*Cyanistes caeruleus*). These birds make up approximately 80 per cent of the Greater Noctule's diet during the birds' migration season in southern Europe. The Greater Noctule is a fast and agile flyer, and can catch birds by fast hawking on the wing; when they are migrating, the birds it eats fly at night. The Neotropical Spectral Bat (*Vampyrum spectrum*) is the largest carnivorous bat and weighs 150–200g (5.3–7.1oz). It eats at least 18 species of bird, as well as large insects, amphibians, reptiles and mammals (including bats), and hunts by slow hawking, locating its prey mainly by scent.

Eating fish is really a specialist kind of carnivory, adopted by around ten species of bat. Only two species are mainly piscivorous, the Greater Bulldog Bat and the Fish-eating Bat (*Myotis vivesi*), which also eats crustaceans and is found only around the Gulf of California, Mexico. Both species also eat insects. Greater Bulldog Bats feed mainly by trawling fish from the surface of the water, after detecting the ripples the fish create. In Europe, the Long-fingered Bat (*Myotis capaccinii*), a species that the IUCN classes as Vulnerable and that is found all around the Mediterranean, eats some fish as well as insects. The three species of sanguivorous vampire bats are also highly specialist; one feeds on the blood of mammals, the other two feed mainly on the blood of birds. They don't kill their prey, so cannot be considered predators.

Below: The Greater False Vampire Bat takes insects as well as larger prey, including other bats. Here, it is preparing to take a rodent.

Bats as food

It can't be easy to catch a fast-moving bat, but bats are energy-rich, attractive food for carnivorous animals. There is no single predator group that takes bats as its main prey, but, worldwide, several animals feed on bats when the opportunity arises.

Wild mammal species, including monkeys, foxes, raccoons, opossums, skunks, weasels, mink, martens, bobcats, other bats, rats and mice eat bats when they can get hold of them. Most of these predators take sick or injured bats, or occasionally healthy ones that are particularly accessible or vulnerable. Humans also eat bats: flying foxes are taken as food, mainly in Asian and Pacific Rim countries and in Africa. In Europe, where bats are protected from human persecution, the main predator of bats is the cat. There are estimated to be around eight million pet cats in the UK, and over 100 million in the whole of Europe. Cats of the feral and domestic varieties can hear ultrasound. Once they discover a bat roost, cats sit outside the entrance and scoop at the bats as they fly in or out. If the roost is in a building, a cat may even be able to enter it and reach the bats.

Some day-flying birds eat bats, such as kites, merlins, hobbies, hawks and falcons, including in particular the Bat

Right: One of the most deadly predators of bats is the domestic cat. Cats often sit outside roosts or in the flight paths of bats, waiting to catch them.

Falcon (*Falco rufigularis*) in Central and South America, and the African Bat Hawk (*Macheiramphus alcinus*). Bat Hawks hunt by waiting next to the mouth of a cave inhabited by bats; one individual was observed in Zambia taking 18 bats per hour. Birds of prey normally only take stray bats flying in daylight, or bats leaving or entering the roost. Bats are particularly vulnerable when they are flying around their roost entrance, as they often have to go through a small hole to get inside the roost and have no escape routes if they are attacked. Crows, jays and other corvids also take bats when they can.

Hibernating bats are not immune from predation. When food is scarce in winter in Eastern Europe, Great Tits (*Parus major*) enter caves and kill and eat hibernating bats; presumably the bats are relatively easy to catch when they are in torpor (see page 22). A pipistrelle must make a huge meal for a Great Tit that normally eats caterpillars.

In summer, active bats share the night sky with nocturnal predators in the form of owls. In the British Isles, bats are

Above: Mexican Free-tailed Bats leaving their roost in Bracken Cave, Texas, USA, are at risk of predation by Red-tailed Hawks.

Right: Some bats are masters of camouflage. These Proboscis Bats, found in Central and South America, are so well-hidden that they can roost on tree bark during the day, often one above another in a vertical line.

estimated to make up less than 0.1 per cent of the diet of owls. There are fewer than 100,000 owls in total in the British Isles, so they are not a huge threat to bats, however, individual owls may develop a taste for bats if they can find a good place to catch them. Tawny Owls (*Strix aluco*) living in the entrance to a large bat hibernation site in Poland were found to eat a diet consisting of up to 40 per cent bats. Tawny Owls weigh 400–800g (14–28oz) and can even tackle big bats such as Noctules and Serotines.

Snakes also eat bats, which they usually catch in caves or trees. Rat snakes, the main culprits, are excellent climbers. They hang down in bat flyways, waiting for bats to come along, or enter roosts in buildings or caves, and kill their prey by constriction. In the New World Tropics, 20 species of snake, mostly boas, are known to eat bats. In Baja California, Mexico, a gopher snake was seen on the wall of an abandoned gold mine, snatching bats from the air as they flew past. It was found to have six bats in its stomach. An African colubrid snake, found in a building, had eight bats and a lizard in its stomach. Various frogs have been seen killing bats, and some lizard and fish species eat bats occasionally. The fish catch bats that are dipping into the water of lakes or rivers to drink. Freshwater Crocodiles (*Crocodylus johnstoni*) use the same strategy and also grab bats that are roosting in vegetation just above the water surface. Saltwater Crocodiles (*Crocodylus porosus*) eat any bats that fall into the water and even jump out of the water to grab bats.

American and Australian Cockroaches (*Periplaneta americana* and *P. australasiae*) eat young and sick bats when they fall to the floor of their roosts. In Caribbean caves, the floor and walls are often covered with cockroaches looking for food; assassin bugs use the same food source. In Venezuela, giant centipedes have been seen eating bats in a cave. Driver ants in Africa can smother young bats and feed on them, while several species of large tropical spider can catch bats in their mist net-like webs, and kill and eat them.

Colonial living does make bats predictable and accessible for some predators, though there is also safety in numbers. Some of the most risky activities for bats are drinking from water inhabited by potentially hungry crocodiles and fish, and entering and leaving roosts. Despite their many predators, bats are pretty good at avoiding being eaten and most manage to live long lives.

Below: Bats that fly low over water or roost near water may be attacked and eaten by crocodiles. This Freshwater Crocodile in Queensland, Australia, is eating a flying fox.

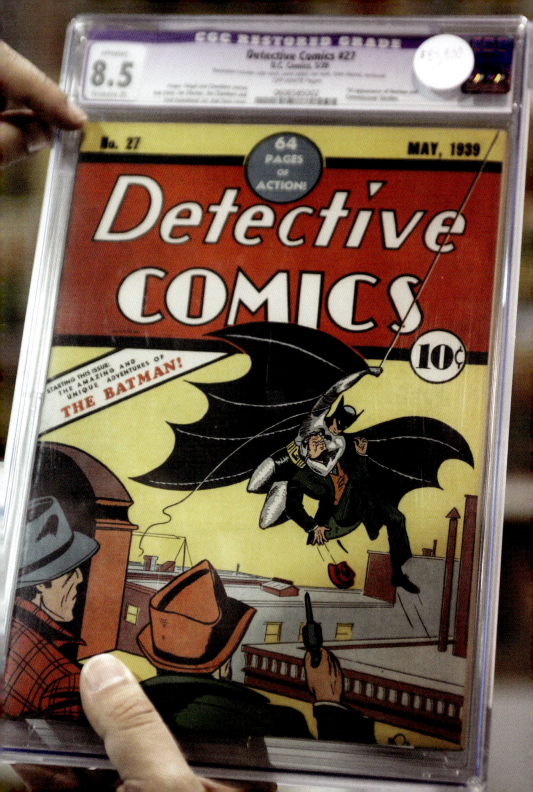

Bats in Culture

Bats are mysterious nocturnal creatures that are often glimpsed but rarely seen well, so it is not surprising that they are the subject of myths and folklore. Depending on your own cultural heritage, you may think of bats as lucky charms, bloodsucking vampires, shapeshifters or witches. Bats have been portrayed as ghosts, spirits and devils, as ugly, scary, dark creatures, as cute, cuddly fluff-balls, and as beautiful, noble and wild things. From comic-book characters to literature and art, this chapter explores the ways in which bats have pervaded our culture.

Bats in folklore

You might have heard that bats get stuck in your hair (extremely unlikely due to their sophisticated echolocation) and are used in magic spells, but did you know that bats are considered lucky by some? In Chinese, the word for bat sounds the same as the word for luck. In Chinese folklore bats represent good fortune, and stylised images of five bats are used to represent the five sources of happiness: wealth, health, long life, virtue and peaceful or natural death.

Opposite: Batman is a fictional superhero who first appeared in DC Comics in 1939. Batman has featured in films, television programmes and comics, and their global success has helped establish the character's prominence in mainstream culture.

Below: Saucers from the Qing dynasty, made in c.1730 in China, showing five bats flying around a peach tree on the front, and more bats on the sides.

Above: This Aztec statue of the bat god Camazotz has been depicted with a nose-leaf.

The Maya people of Central America had a bat god, Zotz (or Camazotz, meaning death bat) who was supposed to occupy the underworld, and who was associated with the night, death and sacrifice. Mayan pictures and carvings of bats created thousands of years ago show that the people were familiar with the bats in their environment; many bats found in Central America today have nose-leaves very similar to the ones shown in the ancient depictions of bats.

A later folk tale from Oaxaca, southern Mexico, a place famous for its traditional colourful bat models and carvings, explains why bats are nocturnal. According to legend, the bat, jealous of the birds, complained to God that it was cold, and God asked the birds to give feathers to the bat to keep it warm. Each bird gave one feather to the bat. The bat, wearing all the different feathers, looked magnificent and brought glorious colour to the sky at night and during the day. But the bat grew arrogant and conceited, and the birds reported its bad behaviour to God. The bat flew across the blue sky and, one by one, the feathers dropped out, leaving the bat in its natural plain state. The bat was so embarrassed that it decided to hide in caves by day and come out only at night to look for its lost feathers. And that is what it has been doing ever since.

Halloween

Bats, perhaps via their connection to witches and ghosts, are today associated with Halloween, an ancient celebration of the end of harvest, the start of winter and death, on 31 October. The origin of Halloween is obscure. It may have Celtic, Pagan or Christian roots, and certainly began in the Western world, though today it is celebrated in many cultures. Halloween marks the start of the darkest part of the year and is seen as a time when spirits are able to enter the world of the living, flitting around like bats. It is also the time when many real bats are beginning to hibernate. Perhaps it is our last chance to see an active bat until spring.

Bats in literature

Left: The three witches or 'weird sisters' in Shakespeare's *Macbeth*, used bat body parts in their spells.

In the play *The Tragedy of Macbeth*, by William Shakespeare, written around 1604, the three witches known as the weird sisters use 'wool of bat' in their brew; this may have helped to establish the cultural link between bats and witches or spells. In folk tales, the wings or blood of bats are also used in spells, or bats themselves are portrayed as witches, ghosts, spirits and supernatural beings or bloodsuckers. In fact, in early European mythology, animals other than bats were associated with bloodsucking and were called vampires. In Caribbean folk stories, the evil ghost-like creature known as the jumbie or duppy is often described as a bloodsucker or vampire, and may be a shape-shifter but is not a bat. Vampire bats in the New World Tropics, when they were described by European scientists after 1800, were named vampires based on the established European folklore.

Below: Count Dracula, here portrayed in the 1931 film by the actor Bela Lugosi, is often shown wearing a bat-like cloak.

From around 1900, bats began to be portrayed as dangerous, terrifying bloodsuckers. The publication of the Gothic novel *Dracula* by Bram Stoker in 1897 was the nail in the coffin for bats. Dracula is able to shape-shift and appears as a large bat, a wolf or even as mist. He leaves a trail of death and other vampires – converted by his bite. Though bats do not feature much in *Dracula*, the connection was made and was strengthened when the story was used in movies and in other books. Since then, bats have been firmly linked to vampires, and vampires are still depicted as bat-like creatures in horror movies and books.

Batman

Batman, a crime-fighting superhero and one of the most famous fictional bat-related characters, was created for 1930s American comic books by Bob Kane and Bill Finger. Batman is the wealthy Bruce Wayne, who swears to fight criminals after witnessing the murder of his parents, and chooses a black bat outfit with a mask and wing-like cape as his scary and intimidating disguise. Wayne lives on the outskirts of the fictional Gotham City in a large mansion with a secret Bat Cave hidden in its foundations. He drives a supercar called the Batmobile, and the authorities can call upon Batman by shining a bat-shaped searchlight, the Bat Signal, into the sky. The comics have stood the test of time and have generated television shows and movies, made with varying degrees of humour and darkness.

Above: The Bat Signal (left) and Batmobile (right) are both featured prominently in the films and comic books.
Below: The figure of Batman has extended beyond his comic book origins and accompanying merchandise is very popular.

Bats in language

Bats have crept into our language a little. 'Like a bat out of hell' means at full speed and 'bat-light' means dusk or darkness. Unfortunately, bats often have negative connotations: 'bat-minded' means mentally blind and an 'old bat' is a foolish woman or girl. Describing someone as 'batty', 'bats' or as having 'bats in the belfry' means they are foolish, odd or crazy. Bats, either real ones or images of bats, are not often found in belfries, but they are found in churches, roosting or carved into the woodwork or stone.

Left: Bats can often be seen carved in the woodwork in churches. This carving is located in the canon's stalls in St Chad's Cathedral, Birmingham, UK.

What's in a name?

In many European languages, the common word for bat includes a reference to mice, showing that bats are identified as small furry things similar to mice but able to fly. An old English name is flittermouse and in other languages we have *vleermuis* (Dutch), *fledermaus* (German), *flädermus* (Swedish), *chauve-souris* (French, meaning bald mouse), *murciélago* (Spanish, meaning blind mouse) and *nahkhiir* (Estonian, meaning leather mouse). The Italian word for bat is *pipistrello*, which is a variant of the Latin *vespertilio*, meaning 'of the evening'. The English word bat may be derived from the Latin *blatta*, meaning moth or nocturnal insect.

Right: These long-eared bats, depicted in an engraving by Lizars, *c*.1800, have been drawn fairly accurately as small, furry, mouse-like creatures that can fly.

Bats in art

Medieval illustrations of bats, like carvings of bats found in churches, are sometimes quite accurate but can also be rather unlike real bats. Instead they show four-legged mammal-like animals with wings on their backs like dragons, or bats consisting of just a head and a huge wing, or even bats with tadpole-like tails instead of hind feet. Artists in medieval times perhaps didn't get to see many bats.

In folk art from Oaxaca, Mexico, brightly coloured bats, other animals and sometimes fantastical hybrids have been made from papier-mâché or carved from wood since the 1930s. These are known as 'alebrije' figures. The bat figures, even the ones that look like skeletons, are lively, engaging and full of character.

Bats have featured in sculptures, heraldry, icons, logos, graffiti, jewellery, textiles, prints and paintings. The shape of a bat lends itself well to stylised art, but bats are also portrayed naturally. The most famous bat in art is perhaps

Right: In this fifteenth century French illustration from *Aesop's Fables*, a bat is shown as neither a mammal nor a bird. In the story, the bat distances himself from both groups, and ends up friendless and alone.

the colour woodblock print by Biho Takashi called *Bat Before the Moon*, made in 1910 and now in the Brooklyn Museum, New York, USA, or the orange-winged bat painted by Vincent van Gogh in 1886 and now in the Van Gogh Museum in Amsterdam, the Netherlands.

Above: In addition to paintings, bats have inspired sculptures, such as this piece entitled *Night Wing* by Dave Whistler, which is located in Austin, Texas, USA.

Left: Biho Takashi's *Bat Before the Moon* is one of the most well-known artworks that depicts a bat.

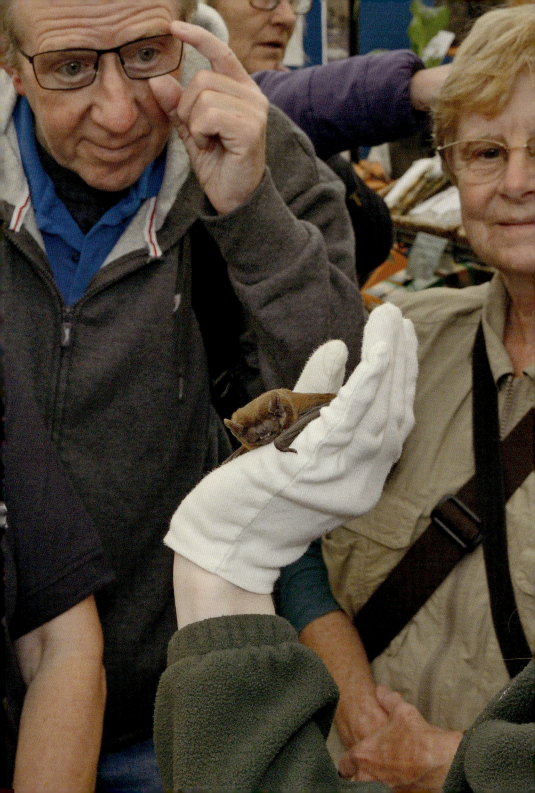

Bats and People

Many people live alongside bats without even noticing them, but numbers of bats have declined in the past due to the activities of humans. Now benefiting from protection throughout Europe, bats are doing better. In the UK, bats cannot legally be harmed or disturbed, impact assessments for planning applications must include consideration of bats by qualified ecologists and scientists studying bats must be licensed for the methods they use. Though bats have been exploited in the past, people and bats are now most likely to interact for research or conservation. Interest in bats has increased, and there are now local bat groups, bat carers and a national charity in the UK, the Bat Conservation Trust.

Inspirational bats

Except where they are taken as food, bats are of little commercial interest to humans. It is lucky for bats that they cannot easily provide enough fur for a coat or enough leather for a pair of gloves. However, bat excrement (guano), collected in vast quantities from cave roosts, has been exploited as a natural fertiliser. Guano was once big business and fortunes were made from its sale; some is still sold today. Over a 20-year period from around 1900, thousands of tons of Mexican Free-tailed Bat guano were removed from Carlsbad Caverns in New Mexico, USA, for sale to fruit farmers in California. At about the same time, guano was being taken from Bracken Cave, Texas; this continued until the late 1980s. Bracken Cave is now owned and protected by Bat Conservation International, so the Mexican Free-tailed Bats that live there are left largely to their own devices.

Opposite: Bats that are being cared for and rehabilitated are sometimes used for education purposes as well. Here, a Noctule is being shown to members of the public at an outreach event in Cornwall, UK.

Below: Bat guano was at one time extracted from roosts in large quantities and used as agricultural fertiliser.

Above: This is a great feeding place for bats: the combination of water and natural vegetation means that insects are plentiful. Habitats like this should be conserved for bats and for other wildlife.

The value of bats as agricultural insect pest controllers is vast but largely unrecognised. Researchers recently measured the value of bats in maize crops in the USA, where the bats not only eat pest insects but also help to control fungi that are spread by insects and damage the crop. The value of bats globally as insect pest controllers in maize crops alone was calculated as over 1 billion US dollars per year. This was worked out from levels of pest-related damage to the maize crop, which were high in experimental plots from which bats were excluded.

During World War II, the USA hatched a plan to use thousands of Mexican Free-tailed Bats to carry time-activated fire bombs. The idea was to exploit the fact that a bat, equipped with a bomb before being released at dawn, has a natural tendency to find a nearby place to roost. The targets were Japanese cities, where most buildings were made of paper and wood and so were highly flammable – and also suitable as bat roosts. Tests were carried out, and during one of these tests armed bats were accidentally released; some of them roosted under a fuel tank and caused fires at the test airfield. In around 1944, the project was abandoned, as the focus switched to the atomic bomb.

More recently, engineers have taken inspiration from the flight of bats. In 2017, robotics experts in the USA unveiled a tiny flying bat-like robot called (obviously) Bat Bot, or B2 for short. It has thin silicon wings stretched over a carbon-fibre skeleton and can fly autonomously. Bat Bot is used to study flapping flight in bats. It may inspire new flight technologies, and is quieter and safer to fly around humans than helicopters and other flying robots because its wings are soft. Presumably, collisions are not too painful.

Even the scary vampire bats have something to offer scientists. Draculin, the anti-coagulant found in their saliva, can be extracted from vampire bats and purified, and is being explored for possible use in medicine. It has potential as a blood-thinning treatment for strokes and heart attacks.

There is still more to learn from bats. Some skilled blind people are able to perceive their environment using echolocation, by making sounds with their mouths, feet, fingers, or with a cane, and listening to the echoes. With training, some can ride bicycles, work out the size of objects and even tell the difference between materials (for example wood and metal) using this method. To 'see' with sound in this way, blind people use the part of the brain normally used for vision. Engineers have developed the ultracane and the ultrabike, mobility devices that use ultrasound to provide information for blind or partially sighted people, allowing them to avoid obstacles. So, being a bat-mimic can help blind people live more active and mobile lives.

Below: Daniel Kish, who has been blind since he was 13 months old, is an American expert in human echolocation. Both Kish and his organisation have taught echolocation to at least 500 blind children around the world. In this photo, he is using his cane in combination with echolocation.

Studying bats

Unless they are conducting research that does not involve handling bats, such as recording echolocation calls or investigating diet from faeces collected without entering a roost, biologists working on bats in the UK are licensed to carry out specific tasks, and must balance any disturbance caused to bats against the knowledge gained.

Early naturalists resorted to the shotgun to obtain bats but found them challenging targets, though fearless and unaware of the danger. More recently, licensed researchers have used hand-nets, mist-nets and harp-traps to catch bats alive. Mist-nets are large nets made from thin nylon thread, which is almost invisible in the dark. Bats that cannot detect mist-nets fly into them and become entangled, after which they are often extremely difficult to untangle. Harp-traps are frames, across which thin fishing line is spanned vertically. Bats that cannot detect the line fly into it and slide downwards into a holding bag. Both nets and traps are usually set at ground level and are not very effective for catching high-flying bats.

Some species of bat are easier to catch than others, and some are very good at escaping once caught – small bats that can hover sometimes fly out of holding bags, and large species, with strong teeth, can bite their way out of mist-nets, leaving holes. Bats normally avoid obstacles by using echolocation, but they do use vision to detect nets and traps that are set in areas lit by lamps or by the moon.

Once a bat has been caught, a wealth of information is available to the keen and curious biologist. Bats can be identified, sexed, weighed and measured, and their wing shape can be quantified. They can be equipped with rings, luminescent tags, radio transmitters or Global Positioning System tags for the study of movements, including migration. Tags are usually attached with a small amount of non-toxic adhesive, such as eyelash glue or colostomy bag glue. Samples can be taken so that scientists can study pathogens and parasites, or for genetic analysis of species and individual relationships. Faeces, pollen and parasites can be collected for study, photographs can be taken and

Below: Harp-traps can be used to collect bats during swarming surveys. Bats that aren't able to detect the thin fishing lines fly into them and slide down into a holding bag, where they can be examined.

Left: This Serotine, caught in the autumn in Wiltshire, UK, had been ringed a year earlier at the same place. Ringing can help improve our understanding of bat populations and the impact of threats that bats face.

echolocation calls can be recorded on release. A few bats can be taken temporarily into the laboratory or placed in a flight cage or wind tunnel for detailed studies of flight, hearing, behaviour or metabolism.

The diet of bats is an important topic for study. Researchers have found out what bats eat by direct observation (which is very difficult in the dark), by analysing the contents of the stomachs of dead bats, and by collecting and analysing bat droppings and food remains dropped by feeding bats.

Feeding bats sometimes drop insect wings or inedible parts of fruits under feeding perches; these food items are usually relatively easy to identify. It is very difficult to identify insects found in the stomachs and droppings of insectivorous bats, as most of them eat small insects that they chew very finely. In the droppings, tiny characteristic pieces of chitin, which makes up the exoskeleton of insects, can be identified under a microscope, but small soft-bodied insects may be underrepresented in relation to bigger ones and moths with lots of scales on their wings. Many pieces are not identifiable. Genetic methods to identify the DNA of insect food from bat droppings are proving useful and will surely provide more detailed information on diet in the future.

The food of frugivorous bats can sometimes be determined from seeds in their droppings. Frugivorous bats also often chew the extract from fruits and then spit out

Right: Prey remains dropped by bats can often be identified to the species level. These butterfly and moth wings were found under a feeding perch used by a Brown Long-eared Bat.

the fibrous remains; researchers can sometimes identify the pulpy pellets the bats discard. Nectarivorous bats, after they have visited several flowers, are often covered in pollen. Researchers have taken samples of this pollen for identification, in order to find out which plants are pollinated by bats, and for genetic studies.

Many questions about bats await answers and researchers are studying their immune systems, disease transmission, genetics, migration, echolocation and evolution. Current 'hot topics' in the conservation of bats relate to agriculture, woodland management, insect populations, wind turbines, street lamps and the ecosystem services that bats provide (such as pest control, pollination and plant dispersal).

Monitoring populations

Counting bats to monitor changes in numbers is pretty challenging, but in most parts of Europe where numbers have been assessed, populations of bats have declined over the last century. Unfortunately, humans are largely to blame, which is unsurprising since the world's human population has quadrupled over the last 100 years. Roosting bats are vulnerable to disturbance by humans and roosts are sometimes destroyed when trees are cut down, mines are filled in or gated, or buildings are renovated or demolished. Bats may also suffer when the habitats that they use for foraging are altered: when

marshes are drained, trees are felled, rivers are canalised or polluted, gardens are paved for parking, hay meadows are converted to silage fields, wild areas are tamed and used for construction, and agriculture becomes more intensive. Insects thrive in diverse habitats, and therefore so do bats, but landscapes are becoming less diverse and insect numbers are declining. Migration and flight routes are also being made more perilous for bats as numbers of wind turbines, power lines and roads are increasing.

Insecticides cause problems for bats, both by reducing the amount of insect food available and by poisoning the bats. Timber treatment in the lofts of houses was a particular problem in the past, but the worst offending products are no longer sold, and all lofts in the UK that are used by bats are protected by law. The use of highly toxic and persistent pesticides such as DDT (dichlorodiphenyltrichloroethane) and lindane (now banned for all but a few purposes in almost all countries) caused huge declines in the numbers of many bat species, not just in Europe but worldwide.

Above: Changing habitats, clockwise from top left: barns like this are ideal roosting places for bats, especially Natterer's Bats, until they are converted into dwellings; river banks with natural vegetation are better for insects and bats than the concrete edges of canals; wind turbines generate electricity in a sustainable way but also kill a lot of bats due to collisions; and mixed farmland with trees, hedges and cattle grazing is good for bats.

Anti-parasitic drugs used to treat cattle remain active in cowpats and lead to reduced numbers of dung beetles for bats to feed on. The food available for bats and insectivorous birds is also reduced by the use of neonicotinoid insecticides (pesticides that are detrimental to non-target insect species, such as bees).

In the UK, monitoring of bat populations is conducted via the Bat Conservation Trust's National Bat Monitoring Programme, a citizen science project involving 3,500 volunteers who carry out four different surveys: hibernation surveys, roost counts, field surveys and waterway surveys. Data collected since 1997 show that numbers of bats are stable, increasing or, in some cases, uncertain. Lesser Horseshoe Bats in particular seem to be thriving. Ongoing monitoring of numbers in roosts and foraging sites is essential for bat conservation, and you can still sign up to contribute to the programme. As well as running the National Bat Monitoring Programme and conducting important research on bat conservation, the Bat Conservation Trust provides evidence-based advice and information to professional ecologists and to members of the public. If you want to put up bat boxes and encourage bats to use your garden, or if you find an injured bat, the Bat Conservation Trust can help.

Right: Bat boxes are useful in woodland where the trees are young and so have few natural cavities. Bats using the boxes are easy for licensed ecologists to count, so the boxes can be an aid to conservation.

Protection

All bat species and their roosts in England and Wales are protected under the Wildlife and Countryside Act (1981) and the Conservation of Habitats and Species Regulations (2010). Similar protection applies in Scotland, Northern Ireland and the Republic of Ireland. In the European Union, bats are protected by the Habitats Directive in order to meet the requirements of the Bern Convention. All 53 European species are also protected under the United Nations Environment Programme's EUROBATS, the Agreement on the Conservation of Populations of European Bats, an international treaty that extends into northern Africa and the Middle East to protect migratory species.

In the UK, it is illegal to capture, injure or kill a bat. It is also illegal to possess, advertise, sell or exchange a bat, or part of a bat, dead or alive. You may also find yourself in serious trouble if you disturb roosting bats, damage or destroy a roost (even when bats are not present) or block an entrance to a roost. In some cases, protection extends to bats' feeding habitats and the flight routes used by bats. Local authorities must consider bats as part of the planning application process. Building work that may affect bats needs to be assessed by a qualified ecologist, so if you may have bats in your house and you are planning any work, you should take advice well beforehand from the Bat Conservation Trust.

Left: Protection of bat roosts often involves placing informative signs and gating mines and caves.

Above: Orphaned bat pups are hand-raised at bat hospitals; here, a bat rehabilitator is giving milk to a Nathusius' Pipistrelle at Kent Bat Group, UK.

The bat hospital

During the summer, baby bats are occasionally separated from their mothers or orphaned, and the lucky ones are cared for by a small but dedicated team of bat rehabilitators. In the UK, these carers can be contacted via the Bat Conservation Trust. Injured or sick adult bats are also taken in, and the bats are released once they are fully recovered and able to fend for themselves. The bats that can't be released, because, for example, they can't fly, are kept in captivity by licensed carers and used for educational purposes. These bats can be taken to schools and events, to allow people to experience a close encounter with a bat. Seeing a bat well in the hand can be an awe-inspiring, inspirational and enchanting experience, especially for a child.

Right: Rehabilitated bats that cannot be released can provide valuable educational experiences. Here, a Noctule is shown to a fascinated audience.

Seeing and hearing bats

It's pretty straightforward to observe bats feeding in summer, but finding and seeing them at roost sites is more difficult. Many bat enthusiasts organise nocturnal bat walks that you can join. And if you are keen, you could buy a bat detector to help you explore the world of bats even more. Being able to hear bats opens up an entirely new way of connecting with them.

Of the wild mammals found in the British Isles, pipistrelles are among the most easily glimpsed, though they are hard to see well. If you go out at dusk at the end of a warm summer's day to a place where insects are flying, you are pretty likely to see bats, and the bats you see are likely to be pipistrelles.

Where and when to look

Habitats that they use for feeding are among the best places to look for bats, as you are most unlikely to disturb them there. Foraging bats will ignore you. They are not normally disturbed by humans with torches, so feel free

Below: You don't have to go far from home to see bats. Here, a Common Pipistrelle flits around trees close to a village in Scotland. Pipistrelles can often be observed hunting at twilight; they emerge from their roosts around half an hour after sunset.

Above: Bats often emerge from their roosts individually or in small groups to avoid predators that may be lurking outside. These male Noctules are flying from a former Black Woodpecker nest hole.

to walk around near rivers or lakes, or in woodland or pasture, on any warm, dry night when insects are likely to be flying. Windy weather is not ideal, and don't bother looking if it is raining or cold, as bats will not fly much when insects are not active. Stick to public footpaths, tracks or lanes, or ask for permission from the landowner. Urban parks can be good places for bats, especially if they feature lakes or ponds.

Dusk and dawn are the best times to see feeding bats, and also to try to find roost sites. At dawn, some species congregate outside their roosts, having a last sociable fly-around before entering; this may help you to find the roost. At dusk, at roost sites, some species can be seen light-sampling: one individual flies out, has a look at the light levels and then flies back in if it is still too light. When light levels are acceptable to the sampling bats, all the bats emerge, but they usually come out individually or in twos or threes to avoid predators that may be waiting by the roost exit. It is easy to imagine the bats lining up inside the roost, the hungry ones that didn't find much food the night before eager to get out but no one wanting to be

the first one out for fear of predators. After a very rainy or cold night all the bats are likely to be hungry, so the next night may be a good time to see them emerge early.

Once you know the location of a nursery roost, you have a perfect place to see bats at dusk, but do be careful not to disturb them. Watch from outside, but don't try to enter the roost or shine a torch at the access point when bats are about to leave.

Above: A Common Pipistrelle (left) leaving its roost in the roof of a barn in Sussex, UK. If you can find a roost, the emerging bats are easy to see. Bat boxes and bird boxes (right) provide roosts for bats and are especially useful in forests where there are few trees old enough to have natural holes and crevices. Here, a Bechstein's Bat leaves its roost in a nest box designed for birds.

Bats in the house?

If you want to know whether or not bats are roosting in your house, choose a warm summer's evening and recruit at least as many friends as you have sides to your house. Each person takes responsibility for one side and watches it closely, preferably while also using a bat detector, as soon as the light starts to fade. The best way to see any bats leaving the building is to lie down in the garden looking up at the house, so that the building is silhouetted against the darkening sky. Keep watching and try not to doze off, as bats are quick and easy to miss. If you and your friends are the kind of people who can get up before dawn, you could try again then, when bats like to spend time flying around near the roost entrance before going in. If there is a roost, what you see will be well worth getting out of bed for.

If you do have bats roosting in your house, consider yourself lucky – you will get some great views of bats coming and going and they are easy guests. Bats don't nibble or chew on wires, pipes or wood like rodents, but they may leave droppings in or near roosts. They don't make holes or nests, or bring bedding or insect food into roosts. In most cases, colonies consist of small numbers of individuals. If you hear noises in your loft at night, it is likely that you have rats or mice, rather than bats, as the bats will be out flying at night and you are likely to hear them only at dusk.

If a bat enters your living space, simply open some windows and doors, and calmly wait for it to leave. In the meantime, you have a perfect opportunity to watch its graceful and beautiful flight. It will soon be on its way.

Below: Dusk and dawn are the best times to see if bats are emerging from or entering your house. A bat detector will help you spot them.

Hitchhiking bats

Bats have not been moved around the globe much by humans, but they have been affected by humans moving predators (including the domestic cat) and food plants around. Bats have been known to 'hitchhike' on boats: a bat finds a nice roost one morning in a crevice in a boat somewhere, and emerges that evening in a completely different harbour. There are records of bats roosting in shipping containers, aircrafts and suitcases. Occasionally bats escape from research facilities or zoos, and end up having to explore and inhabit a new area.

Bat or bird?

Against a dark sky at the end of the day, a small dark shape flits past. What was it? It could have been a Blue Tit or a House Sparrow (*Passer domesticus*), heading towards its roosting place, or a Swallow or Swift, staying up late to feed on the dusk-flying insects. Or it could have been a bat, just emerged and ready for the night ahead. The small birds that can be confused with bats are getting ready for their resting period at dusk, and if you watch them for a few seconds they tend to land somewhere. Bats tend not to land – many species, including pipistrelles, catch insects that are small enough to eat on the wing. Long-eared Bats and Greater Horseshoe Bats, when feeding on larger moths or beetles, land in order to process their catch. But even these species, if watched when in flight, are likely to fly continuously rather than from perch to perch like birds.

Organised bat walks

Look out for bat walks or other bat-related events organised in the UK by the Bat Conservation Trust, your local bat group, park rangers, the National Trust, your Wildlife Trust or other wildlife charities. Bat walks are a great, safe way to find out about good places to see and hear bats, and the organisers usually have bat detectors for you to try out. You will learn about the different species found in your area and perhaps you will be inspired to learn more about bats, or even to buy your own detector. The long, dark night will never be the same again once you discover the secretive but fascinating world of bats.

Above: Simple bat detectors are fun to use and make it much easier to see bats, and a bat walk is a great excuse to stay up late.

Below: Members of Kent Bat Group surveying chalk caves, known as deneholes, for hibernating bats. Bat groups welcome new members and often organise bat walks.

Above: Noctules are tree-dwellers, preferring to roost in old woodpecker holes or other crevices in trees. They are usually the first bat species to emerge in the evenings and their loud calls can be detected at over 100m (328ft) away.

Glossary

Carnivorous Feeding on the flesh of other animals (e.g. bats, other mammals, reptiles, frogs).

Colony A group of individuals (e.g. bats) living together in close association and forming a social unit.

Cryptic species Very similar species that are difficult to identify, such as Common and Soprano Pipistrelles.

Echolocation Also called biosonar. A sensory system used by bats to collect information about the environment by producing calls and listening to the echoes bouncing back from objects.

Frugivorous Feeding on fruit. Frugivorous bats often aid in the dispersal of plants, by moving the seeds away from the fruiting tree. The seeds are dropped or spat out, or pass through the digestive tract and are deposited with a small amount of guano.

Gleaning Feeding by removing prey from a surface (e.g. a leaf or the ground). Many gleaning bats are able to hover.

Guano Bat faeces, often called guano when collected from roosts and used as fertiliser.

Hawking Capturing prey while in flight. Bats feeding by hawking usually eat their small prey while on the wing. Hawking flight can be fast or slow.

Hibernation Prolonged periods of torpor, used during winter to conserve energy.

Insectivorous Eating insects and/or other invertebrates (arthropods, e.g. spiders).

Nectarivorous Feeding on nectar. Nectarivorous bats often also eat pollen, and pollinate flowers by carrying pollen on their fur from plant to plant.

Nose-leaf A fleshy leaf-like structure on the nose of some bats, used in echolocation.

Nursery roost or **maternity roost** Place where a colony of female bats gathers in daytime in summer to give birth. Nursery roosts are occupied from around May, most bats in Europe give birth in June, and the young bats start to fly in July and August. Then the nursery colonies separate.

Omnivorous Eating many different kinds of food. For example, many bats eat insects as well as fruit.

Perch-hunting Capturing prey during a short flight from a vantage point, then returning to the perch to eat the prey.

Piscivorous Fish-eating.

Roost A place inhabited by a colony of bats or by an individual bat.

Sanguivorous Feeding on blood.

Songflight A territorial display flight that involves song, carried out by Skylarks (*Alauda arvensis*) and other birds. Male pipistrelles and other bat species also perform songflights by flying or gliding in loops while making special advertisement calls in August and September to attract females for mating.

Torpor Lowering the body temperature in a controlled way, in order to reduce the amount of energy needed.

Trawling In bats, feeding by capturing prey from the water surface, by dragging the feet or tail membrane through the water.

Further Reading and Resources

Books

Jones, Kate and Walsh, Allyson. *A Guide to British Bats*. Field Studies Council, 2001.

Dietz, Christian and Kiefer, Andreas. *Bats of Britain and Europe*. Bloomsbury, 2016.

Fenton, Melville and Simmons, Nancy. *Bats: a World of Science and Mystery*. University of Chicago Press, 2015.

Russ, Jon. *British Bat Calls: A Guide to Species Identification*. Pelagic Publishing, 2012.

Online

Bat Conservation International An organisation based in Texas, USA, devoted to conserving the world's bats and their ecosystems: batcon.org

Bat Conservation Trust A UK non-governmental organisation solely devoted to the conservation of bats and their habitats. The website provides advice about how to deal with any bat-related problems or queries, how to contact your local bat group, how to join the National Bat Monitoring Programme and how to find a bat carer: bats.org.uk; 0345 1300 228

BatLife Europe An international umbrella organisation for partner bat conservation organisations throughout Europe, such as the Bat Conservation Trust: batlife-europe.info

Bats and Churches Partnership A partnership of Natural England, Church of England, Historic England, Bat Conservation Trust and Churches Conservation Trust, working together in the UK to support churches with bats: batsandchurches.org.uk

Bats in Flight Guide A useful guide produced by the Natural History Museum, London, to help people identify bats in flight: bit.ly/2I6UoPv

National Trust There are lots of bat roosts at National Trust properties, and some have webcams. The National Trust also organises bat walks: nationaltrust.org.uk

Vincent Wildlife Trust A wildlife charity specialising in conservation of bats and other animals, which owns and manages several horseshoe bat roosts as reserves: vwt.org.uk

Wildlife Trusts An umbrella organisation for all the local Wildlife Trusts, many of which organise bat walks and other bat-related events: wildlifetrusts.org

Acknowledgements

At Bloomsbury, I thank Jenny Campbell, Julie Bailey and Jim Martin, who originally suggested me as a Spotlight author. Louise Morris did a brilliant job as my copy-editor. I thank my friends who kindly made suggestions, provided information and corrected some of my errors: Annika Binet, Gareth Jones, Joe Nunez-Mino, Kirsty Park, Stuart Parsons, Danilo Russo, Shirley Thompson, Dean Waters and Lisa Worledge. Adam Britton told me about crocodiles eating bats, and Margaret Vaughan read and improved the manuscript.

IMAGE CREDITS

Image Credits

Bloomsbury Publishing would like to thank the following for providing photographs and permission to reproduce copyright material. While every effort has been made to trace and acknowledge all copyright holders, we would like to apologise for any errors or omissions and invite readers to inform us so that corrections can be made in any future editions of the book.

Key t = top; l = left; r = right; tl = top left; tcl = top centre left; tc = top centre; tcr = top centre right; tr = top right; cl = centre left; c = centre; cr = centre right; b = bottom; bl = bottom left; bcl = bottom centre left; bc = bottom centre; bcr = bottom centre right; br = bottom right

AL = Alamy; FL = FLPA; G = Getty Images; IS = iStock; NPL = Nature Picture Library; RS = RSPB Images; SH = Shutterstock

Front cover t Eric Medard/NPL, b Buiten-Beeld/AL; **spine** Paul Sawer/FL; **back cover** t MYN/Paul van Hoof/NPL, b Nick Upton (rspb-images.com)/RS; **1** Michael Durham/Minden Pictures/G; **3** David Hosking/FL; **4** Simon Colmer/NPL; **5** De Agostini Picture Library/G; **6** t Michael Durham/NPL, b Kim Taylor/Nature Picture Library/G; **7** Jiri Lochman/NPL; **8** tl Rick & Nora Bowers/AL, tr Michael Durham/NPL; **9** Photo Researchers/FL; **10** MShieldsPhotos/AL; **11** Tim Laman/G; **12** t ©Lizzie Harper, b Stephen Dalton/Minden Pictures/G; **13** Michael Durham/Minden Pictures/G; **14** Dietmar Nill/Nature Picture Library/G; **15** Chris Howes/Wild Places Photography/AL; **16** Peter Verhoog/Buiten-beeld/Minden Pictures/G; **18** Lucas Oleniuk/Contributor/G; **19** Michael Durham/Minden Pictures/G; **20** Jose B. Ruiz/NPL; **21** Dietmar Nill/NPL; **22** Zolran/SH; **23** Paul van Hoof/Buiten-beeld/Minden Pictures/G; **24** belizar/SH; **25** Frank Greenaway/G; **26** bl Arterra/Contributor/G, br Stephen Dalton/Minden Pictures/G; **27** bl Mantonature/IS, br All-stock-photos/SH; **28** Paul van Hoof/Buiten-beeld/Minden Pictures/G; **29** t Nature Photographers Ltd/AL, b Paul van Hoof/Buiten-beeld/Minden Pictures/G; **30** Dietmar Nill/NPL; **31** t Simon Colmer/NPL, b Karl Van Ginderderuen/Buiten-beeld/Minden Pictures/G; **32** Hugo Willocx/Biosphoto/FL; **33** Paul van Hoof/Buiten-beeld/Minden Pictures/G; **34** Stephen Dalton/Minden Pictures/G; **35** t imageBROKER/AL, b Hugh Clark/FL; **36** t Nature Photographers Ltd/AL, b Paul van Hoof/Buiten-beeld/Minden Pictures/G; **37** Paul van Hoof/Buiten-beeld/Minden Pictures/G; **38** Stephen Dalton/NPL; **40** Nick Upton/NPL; **41** Mike Lane (rspb-images.com)/RS; **42** t Nick Upton/NPL, b Belizar/SH; **43** Nick Upton/NPL; **44** t skapuka/SH, bl Michael Pesata/SH, br Simon Colmer/NPL; **45** salajean/SH; **46** tc Michael Pesata/SH, tr Alis Photo/SH, cl KOO/SH, c Gucio_55/SH, cr David Dohnal/SH, bl neil hardwick/SH, bc Simon Colmer/NPL, br David Dohnal/SH; **47** tc neil hardwick/SH, tr blickwinkel/AL, cl David Dohnal/SH, c Stephen Farhall/SH, cr Jean-Franois Noblet/G, bl Dietmar Nill/NPL, bc Nature Photographers Ltd/AL, br Nick Upton/NPL; **50** Photo Researchers/FL; **51** Dale Sutton (rspb-images.com)/RS; **52** Dietmar Nill/NPL; **53** Colin Underhill/AL; **54** Marcin Perkowski/SH; **55** Ann and Steve Toon/AL; **56** Arterra/Contributor/G; **57** Jim Clark/AL; **58** Konrad Wothe/NPL; **59** DESIGNFACTS/SH; **60** t Christian Ziegler/Minden Pictures/G, b Christian Ziegler/Minden Pictures/G; **61** Christian Ziegler/Minden Pictures/G; **62** Nick Hawkins/NPL; **63** Nicolas Reusen/G; **64** Minden Pictures/AL; **65** Phil Savoie/NPL; **66** Michael Durham/Minden Pictures/G; **67** Rolf Nussbaumer/NPL; **68** Christian Ziegler/Minden Pictures/G; **69** FLPA/AL; **70** Dietmar Nill/Minden Pictures/FL; **71** Arco Images GmbH/AL; **72** Minden Pictures/AL; **73** Bloomberg/Contributor/G; **74** tl Piotr Naskrecki/Minden Pictures/G, tr Jiri Prochazka/SH, b Nick Upton/NPL; **75** Gerard Lacz/FL; **76** Kerstin Hinze/NPL; **77** Photononstop/AL; **78** Richard Wayman/AL; **79** Kerstin Hinze/NPL; **80** Juan Carlos Munoz/G; **81** Hugh Clark/FL; **82** Dietmar Nill/NPL; **83** Dietmar Nill/NPL; **84** Stephen Dalton/NPL; **85** Photo Researchers/FL; **86** Kim Taylor/NPL; **87** Minden Pictures/AL; **88** MoMorad/IS; **89** tl David Chapman/AL, tr Marek R. Swadzba/SH, cr Lee Hua Ming/SH, br Gucio_55/SH; **90** Ch'ien Lee/Minden Pictures/G; **91** Melinda Fawver/SH; **92** tl Vishnevskiy Vasily/SH, tr Dietmar Nill/NPL; **93** Stephen Dalton/Minden Pictures/G; **94** Raniel Jose Castaeda/EyeEm/G; **95** Philip Dalton/NPL; **96** Bill Gozansky/AL; **97** Cathy Finch/G; **98** Spencer Weiner/Contributor/G; **99** Heritage Images/Contributor/G; **100** t Print Collector/Contributor/G, b Nick Ballon/G; **101** t Wikimedia Commons, b World History Archive/AL; **102** tl Albert L. Ortega/Contributor/G, tr Bettmann/Contributor/G, b Bloomberg/Contributor/G; **103** t Alex Ramsay/AL, b UniversalImagesGroup/Contributor/G; **104** De Agostini/J. L. Charmet/G; **105** t Peter Tsai Photography/AL; b Gift of the Estate of Dr. Eleanor Z. Wallace/Brooklyn Museum; **106** Nick Upton/NPL; **108** Arkadiusz Warguła/IS; **109** ZUMA Press, Inc./AL; **110** Nick Upton/NPL; **111** Nick Upton/NPL; **112** Stephen Dalton/NPL; **113** tl Andrew Roland/SH, tr lgabriela/SH, cl Matthew Dixon/SH, cr Steve Meese/SH; **114** Mike Powles/SH; **115** The National Trust Photolibrary/AL; **116** t Terry Whittaker/2020VISION/NPL, b Nick Upton/NPL; **117** Laurie Campbell/NPL; **118** Klaus Echle/NPL; **119** tl Stephen Dalton/NPL, tr blickwinkel/AL; **120** Terry Whittaker/2020VISION/NPL; **121** Dale Sutton (rspb-images.com)/RS; **122** t Manor Photography/AL, b Terry Whittaker/2020VISION/NPL; **123** David Chapman/AL.

Index